T0140950

Lignin as an Alternative Precursor for a Sustainable and Cost-Effective Carbon Fiber for the Automotive Industry

Vom Fachbereich Produktionstechnik
der
UNIVERSITÄT BREMEN

zur Erlangung des Grades
Doktor-Ingenieur
genehmigte

Dissertation
von

Diplomingenieur Hendrik Mainka

Gutachter:
Prof. Dr.-Ing. Axel S. Herrmann (Universität Bremen)
Prof. Dr. Tim A. Oswald (University of Wisconsin-Madison)

Tag der mündlichen Prüfung:
30.03.2015

Science-Report aus dem Faserinstitut Bremen

Hrsg.: Prof. Dr.-Ing. Axel S. Herrmann

ISSN 1611-3861

Die Ergebnisse, Meinungen und Schlüsse dieser Dissertation sind nicht notwendigerweise die der Volkswagen AG.

Bibliografische Information der Deutschen Nationalbibliothek

Die Deutsche Nationalbibliothek verzeichnet diese Publikation in der Deutschen Nationalbibliografie; detaillierte bibliografische Daten sind im Internet über http://dnb.d-nb.de abrufbar.

©Copyright Logos Verlag Berlin GmbH 2014

Alle Rechte vorbehalten.

ISBN 978-3-8325-3972-6

Logos Verlag Berlin GmbH
Comeniushof, Gubener Str. 47,
10243 Berlin
Tel.: +49 030 42 85 10 90
Fax: +49 030 42 85 10 92
INTERNET: http://www.logos-verlag.de

To my mother and father with love

Acknowledgements

I would like to thank all the individuals and institutions for their kind support and co-operation by providing me with valuable information and feedback during the creation of this dissertation. Without this help and support this dissertation would have never reach this stage.

My great acknowledgement goes to Prof. Dr. Axel S. Hermann from the Fiber Institute Bremen at the University of Bremen (Germany) for giving me the opportunity to work with his group, and for being a great supervisor throughout this research and during the creation of this dissertation.

Most of all I would like to thank the Volkswagen AG for welcoming me as a PhD candidate into the Group Research in Wolfsburg (Germany) and allowing me to enjoy its wonderful research opportunities.

I am very thankful that Dr. Armin Plath and Dr. Olaf Täger gave me the opportunity to be part of the Polymer Group at the Volkswagen Group Research in Wolfsburg.

Special thanks go to Mr. Robert E. Norris, Jr. for giving me the opportunity to research with the Carbon Materials Technology Group at the Oak Ridge National Laboratory in Oak Ridge Tennessee (United States of America). I also express my heartfelt gratitude to Mr. Fue Xiong for his help in all laboratory and administrative work at the Oak Ridge National Laboratory and his inspiring discussions.

Special thanks go to Dr. Enrico Körner and Mr. Oliver Stoll from Volkswagen AG, Wolfsburg for their valuable suggestions and beneficial discussions.

A big thank you goes to Dr. Liane Hilfert for her support concerning nuclear magnetic resonance spectroscopy and infrared spectroscopy, and many helpful ideas.

I take this opportunity to also thank Dr. Thorsten Städler for the Raman spectroscopy measurements and the pictures from the scanning electron microscope, and Dr. Patrick Rupper for his help concerning the X-ray photoelectron spectroscopy.

I am indebted to my thesis committee, Prof. Dr. Tim A. Oswald from the Department of Mechanical Engineering of the University of Wisconsin in Madison, and Prof. Dr. Prof. Dr. rer. nat. Bernd Mayer from the Fraunhofer Institute for Manufacturing Technology and Applied Materials Research Bremen, as well as Dr. rer. nat. Thorsten Staedler from the Department Surface and Materials Technology of the university of Siegen who kindly agreed to be co-examiners in the colloquium.

I would also like to extend my deepest gratitude to all other colleagues in the Polymer Group of Volkswagen Group Research, as well in the Carbon Materials Technology Group at the Oak Ridge National Laboratory, for all of the interesting and successful meetings and discussions, and for the wonderful years we spent together.

Last but not least, I would like to thank my parents for their love and encouragement.

Kurzfassung

Leichtbau ist bei Volkswagen integraler Baustein einer ganzheitlichen CO_2-Reduzierungsstrategie. Ein geringeres Fahrzeuggewicht steht in direkter Korrelation zur Reduzierung der CO_2-Emmission beim Betrieb des Fahrzeuges. Kohlenstofffaserverstärkte Kunststoffe (CFK) bieten die Möglichkeit bei gleicher Funktionalität wie Stahlbauteile eine Gewichtsreduktion von bis zu 70 % zu erzielen, und damit CO_2-Emmissionen signifikant zu senken. Allerdings werden die Gewichtsvorteile durch die bisher hohen Herstellungskosten für die Kohlenstofffasern relativiert. Bei der Betrachtung der Herstellungskosten wird deutlich, dass über 50% der anfallenden Kosten auf den Precursor (Ausgangsprodukt) entfallen.
Alternative Precursoren bieten die Möglichkeit, die Herstellungskosten für Kohlenstofffasern erheblich zu senken. Die wichtigsten Kriterien bei der Auswahl eines Precursors sind der Preis, die Verfügbarkeit und die ökologische Nachhaltigkeit.
Lignin erfüllt als ein Abfallprodukt, welches bei der Papierherstellung in enormen Mengen anfällt, diese Kriterien am besten.
Da es sich bei Lignin um ein Naturprodukt handelt, ergeben sich abhängig vom Isolationsprozess und der verarbeiteten pflanzlichen Rohstoffe unterschiedliche Eigenschaften der gewonnenen Lignine. Um genaue Vorstellung von Lignin als Precursors für die Herstellung von Kohlenstofffasern zu erlangen, wurden dessen Eigenschaften, Zusammensetzung und chemische Struktur systematisch bestimmt und aufgeklärt.
Für die Herstellung einer Kohlenstofffaser aus Lignin im Labor-Maßstab wurden die wesentlichen Prozessschritte etabliert.
Um ein genaueres Verständnis der während des Produktionsprozesses ablaufenden chemischen Reaktionen zu erlangen, wurden die Eigenschaften und die chemische Struktur der Lignin-Produkte entlang der Prozesskette systematisch untersucht. In dieser Dissertation wird erstmalig die Herstellung einer Kohlenstofffaser auf Basis von Lignin, vom Precursor über den Konversionsprozess bis hin zur Kohlenstofffaser, betrachtet und das Potential von Lignin als alternativer, nachhaltiger und kostengünstiger Precursor bestimmt.

Abstract

Lightweight construction is an integral part of the overall Volkswagen strategy for the reduction of CO_2 emissions. A lower vehicle weight has a direct correlation to the reduction of the CO_2 emissions. One potential way to reduce weight is the utilization of lightweight carbon fiber reinforced plastics (CFRP). When comparing CFRP to steel, a mass reduction of up to 70% in automobile parts without a degradation of the mechanical properties is possible. This mass reduction leads to a significant saving of CO_2 emissions.

Currently the drawbacks of using CFRPs are the high manufacturing cost. This is due to the fact that over 50% of the cost of conventional polyacrylonitrile (PAN) based carbon fiber, belongs to the manufacturing cost of the precursor. The manufacturing cost of carbon fiber can be significantly reduced by the use of alternative precursors. The decision to use an alternative precursor to produce carbon fiber is based on the price, the availability, and the renewability of the precursor in question.

Lignin is a natural waste byproduct of the paper industry, which is readily available in enormous amounts and due to this very inexpensive. Since lignin is a natural product it can have different properties depending on the plant in which it is derived from and the process in which it is isolated. To get a better understanding of lignin as a precursor for carbon fiber production, a systematical analysis of the properties, compounding, and chemical structure is made.

To investigate the potential of lignin as an alternative precursor a set of fundamental production steps were needed to be established, since there was no procedure to produce carbon fiber from lignin. These steps consist of: pelletizing, fiber spinning, fiber stabilization, and carbonization. The intermediate products of these steps are the lignin powder, the lignin pellets, the lignin fiber as well as the oxidized and carbonized lignin fiber. At each step the properties and chemical structure were analyzed.

This dissertation shows for the first time in research the complete production process of carbon fiber made from lignin by analyzing the chemical structure of all intermediate products, and determines the potential of lignin as an alternative, economic and ecological precursor.

Table of Content

List of Figures

List of Tables

1 Introduction

As of January 1st 2014, all new vehicles are required to meet Euro 6 emission standard. This standard requires all new light passengers and light commercial automotive manufactures to reach an average vehicle Carbon Dioxide (CO_2) emission of 130 g/km. The average vehicle emission of a light passenger vehicle produced by the brand Volkswagen in 2012 was 130 g/km. The figure below shows the German CO_2 emission and the amount needed to be reduced for the coming years (Figure 1).

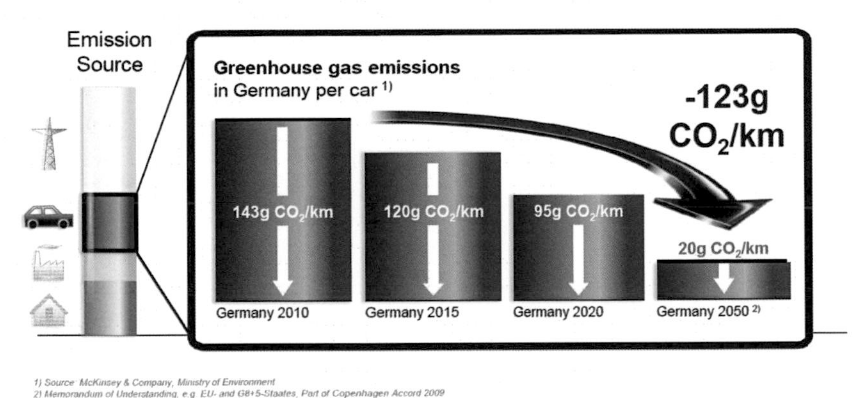

Figure 1: CO2 emission standards in Germany [1-4]

The emission standards for 2020 have already been determined by the European Union (EU) and by the United States of America Environment Protection Agency (EPA). The next set of regulations been set for both Euro 7 and Tier 3 with an average light weight vehicle emission of CO_2 of 95 g/km and 114 g/km, respectively.

The standards for the CO_2 emissions in 2020 are shown in Table 1.

Table 1: Emission standards for 2020 [1-1, 1-2, 1-3]

	Limit value for carbon dioxide emissions in 2020
EU	95 gCO_2/km
USA	114 gCO_2/km
Japan	113 gCO_2/km

These new emission standards provide a vast amount of challenges for the automotive manufactures. A significant reduction in their vehicle's emissions is necessary to prevent potential fines threated by the government.

To meet these onerous regulations, manufactures must focus their resources. For several years the focus was on more efficient powertrains and drivetrains but it is now clear that these emission standards will not be achieve by improvements solely on this. Manufactures must expand their research into fields such as aerodynamics in hopes to reduce overall vehicle drag, and the usage of advance lightweight materials to reduce the overall vehicle's weight.

The average vehicle weight has been increasing over the past decades. Higher safety standards, better quality, more interior and exterior conveniences have led to this rise in weight (Figure 2). Almost every car generation has increased in weight. Stronger chassis members, such as the side impact bars, needed to meet increasing safety standards, which led to heavier chassis. To archive the same dynamic performance, the car must be equipped with a heavier high performance engine with higher fuel consumption. This leads to bigger heavier fuel tank. This is one example of how the weight spiral occurs. The principle of this "weight spiral" shows this dilemma (Figure2) [1-4].

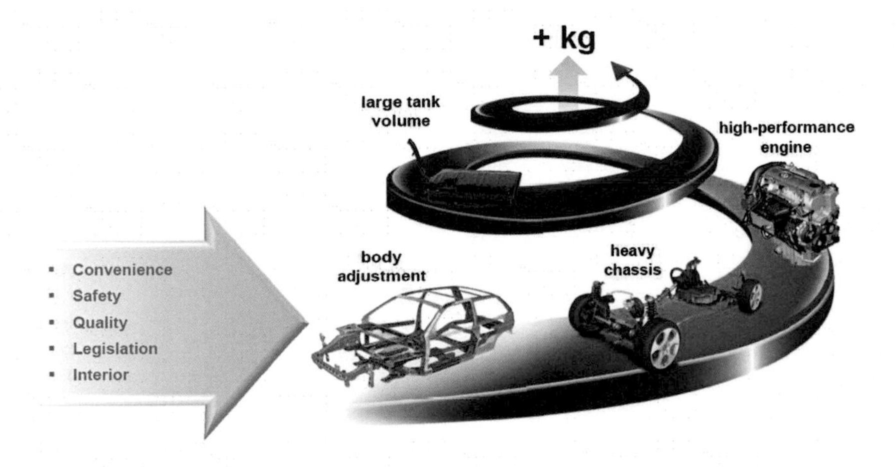

Figure 2: Principle of "the weight spiral" [1-4]

For automotive manufactures to reach the 95 gCO_2/kg requirement by 2020, the implementation of lightweight constructions can not solely be based on steel (compare Volkswagen Golf) and aluminum (compare Audi A8) advancements. To accomplish this, a reduction of component weight must exceed 50%. Figure 3 shows the potential of weight saving per each lightweight construction material. Steel and aluminum can only attribute up to 40% weight savings [1-4].

Figure 3 shows the potential of lightweight construction carbon fiber reinforced plastics in structural applications and the related cost factors, which would be realized by substitution of a steel component by carbon fiber reinforced plastics [1-4].

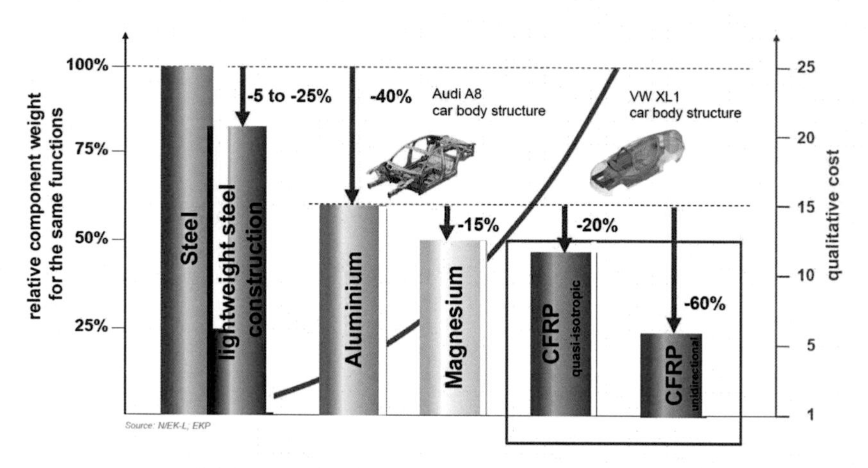

Figure 3: Potential of lightweight construction - material substitution in structural applications [1-4]

Carbon fiber reinforced plastics (CFRP) can provide this 50% minimum reduction in weight. As Figure 3 shows, a weight saving potential of up to 70% on a component without compromising its functionality is possible. The ability to reverse the "Weight Spiral" is finally possible with the use of advanced CFRPs and their implementation in the automotive industry. Figure 4 shows how this will affect the "Weight Spiral" [1-5]. The reversing of the "weight spiral" is finally necessary to reach the future emissions standards. How the reversing of the "weight spiral" could work is shown in Figure 4 [1-4].

Lighter body adjustments lead to lighter chassis which require less power such that downsizing of the powertrain and drivetrain occurs. This in turn requires less fuel which reduces the fuel capacity required. All of which reduce the vehicles overall weight (Figure4).

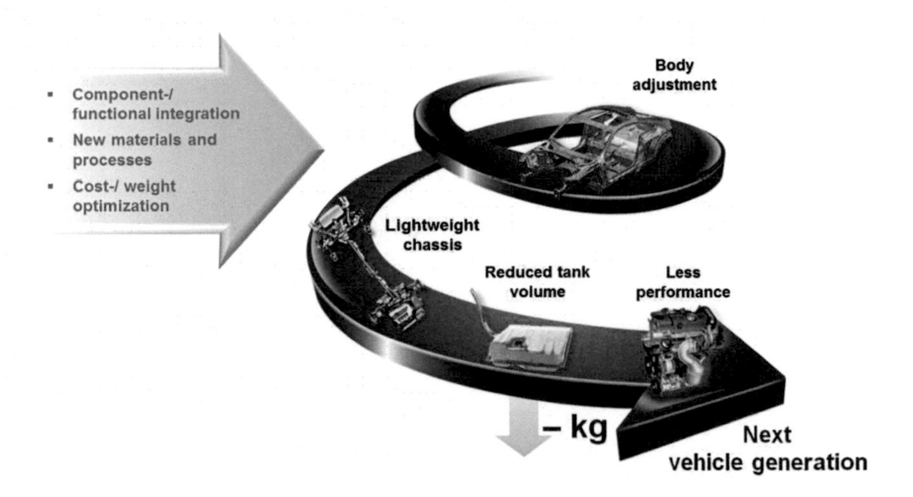

Figure 4: Reversing of "the weight spiral" [1-4]

For customers of the automotive industry lightweight construction has enormous benefits. In context the reduction of fuel consumption, the increase of the range, as well as a better driving dynamics of the vehicle are the main benefits of lightweight construction using carbon fiber reinforced plastics. For example, the affects of a lightweight construction that reduces the weight of a car by 100 kg, means for the standard engine, a reduction of fuel consumption of 0.3 l/100 km. For an electric drivetrain this would result in a range increase of 100 km [1-6, 1-7].

However, the weight benefits are currently limited by the high production cost of the carbon fiber. In Figure 3 the cost factors which are connected to the different materials is illustrated by the blue line. The cost factors are not to be disregarded and one possibility to reduce the cost of carbon fiber is to change the precursor used to produce carbon fiber.

Today the most common precursor used to produce carbon fiber is polyacrylonitrile (PAN). When investigating the cost of producing carbon fiber, it becomes clear that more than 50% of the cost is related to the production of the precursor. The remaining cost comes is distributed as such; 15% to the oxidation process, 23% to the carbonization process, and the remaining to sizing and spooling (Figure 5) [1-8].

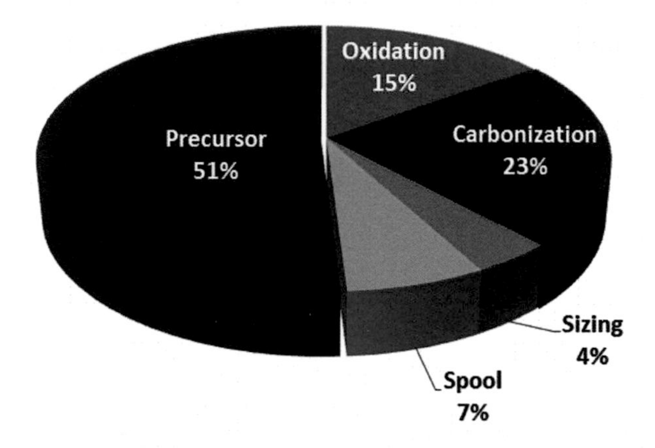

Figure 5: Cost distribution for the production of PAN based carbon fiber [1-4]

The key to reducing the production cost of carbon fiber is the use of an alternative precursor. The alternative precursors with the highest potential for the automotive industry are polyethylene and lignin.

If lignin is compared to polyacrylonitrile as a precursor for carbon fiber, the ability to reduce the cost of manufacturing is by more than 50% [1-9]. That's the reason for the investigation a lignin based carbon fiber in this thesis. Lignin is a natural waste by-product of the paper industry and biorefineries, which is readily available in enormous amounts and due to this very inexpensive. Lignin is also a sustainable, renewable resource. The use of lignin offers significant cost saving potential in the production of carbon fiber. Lignin makes it possible to produce a carbon fiber based on renewable resources.

1.1 Scope of the Dissertation

The work presented in this dissertation defines the process to produce carbon fiber from lignin, and the analysis of the properties and chemical structure of lignin as a precursor material. The properties of the fibers produced during the production process of lignin based carbon fiber are also analyzed. For the evaluation of the properties and chemical structure many innovative techniques were used including thermo gravimetric analysis, differential scanning calorimetry, gas pycnometry, elementary analysis, mass spectroscopy, nuclear magnetic resonance spectroscopy, Fourier transform infrared spectroscopy, Raman spectroscopy and X-ray photoelectron spectroscopy.

A lignin based carbon fiber has been produced on a laboratory scale as well as a pilot line scale. This thesis shows the potential that lignin has as a precursor for carbon fiber production, and the investigation of the reaction mechanism of the conversion process.

1.2 Structure of the work

The chapters are organized as follows: Chapters 2 and 3 describe industrial standards for producing carbon fiber. Chapter 2 describes the carbon fiber production process using the PAN based carbon fiber as an example, beginning with the precursor production followed by stabilization and carbonization process, as well as surface treatment and sizing. Chapter 3 explains the advantages and disadvantages of different precursors used to produce carbon fiber.

Chapters 3 and 4 specify lignin as a precursor for carbon fiber. Therefore different isolation processes from wood are presented and the properties and chemical structure of the used Hardwood Lignin is investigated.

Chapter 5 presents, as one of the main parts of the thesis, the production process of a lignin based carbon fiber with the steps: compounding and pelletizing, precursor fiber production, fiber stabilization, and fiber carbonization.

The reaction mechanism of lignin based carbon fiber is proposed in Chapter 6. This chapter helps to understand the chemical changes during conversion process of lignin to carbon fiber.

Chapter 7 is used to define the properties of a lignin based carbon fiber using techniques like single fiber tensile test, gas pycnometry and scanning electron microscopy. Chapter 7 also shows the investigation of the chemical structure of lignin fiber, stabilized and carbonized lignin based fibers by Raman spectroscopy, X-ray photoelectron spectroscopy and secondary ion mass spectrometry.

A comparison between the new lignin based carbon fiber and a conventional PAN based carbon fiber using Raman spectroscopy, X-ray photoelectron spectroscopy and scanning electron microscopy is given in Chapter 8.

In Chapter 9 the potential of lignin based carbon fiber is summarized and an outlook for future development is given, followed by the concluding remarks.

1.3 References

[1-1] Verordnung (EG) Nr. 443/2009 des europäischen Parlamentes und Rates vom 23. April 2009 zur Festsetzung von Emissionsnormen für Personenkraft wagen im Rahmen des Gesamtkonzepts der Gemeinschaft zur Verringerung der CO2-Emissionen von Personenkraftwagen und leichten Nutzfahrzeugen.

[1-2] Federal Register USA – Department of Transportation - Pages 62623-63200 Vol. 77, No. 199 / Monday, October 15, 2012 / Rules and Regulations 2017 and Later Model Year Light-Duty Vehicle Greenhouse Gas Emissions and Corporate Average Fuel Economy Standards

[1-3] Japan Automobile Manufacturers Association, November 2011 2011 Report on Enviromental Protection Efforts

[1-4] H. Mainka, O. Täger, O. Stoll, E. Körner, A. S. Herrmann: Alternative

Precursors for Sustainable and Cost-Effective Carbon Fibers usable within the Automotive Industry.
Society of Plastics Engineers (Automobile Division) – Automotive Composites Conference & Exhibition 2013, Novi, Mich. USA

[1-5] T. Suzuki et al.: LCA for lightweight vehicles by using CFRP for mass-produced vehicles Proceedings of the 15th International Conference on Composite Materials (JCCM15) ,2005

[1-6] A. Mayyasa, A. Qattawia, M. Omara, D. Shana Design for sustainability in automotive industry: A comprehensive review Renewable and Sustainable Energy Reviews Volume 16, Issue 4, May 2012, Pages 1845–18623

[1-7] R. Witik, J. Payet, V. Michaud, C. Ludwig, J.-A. Månson
Assessing the life cycle costs and environmental performance of lightweight materials in automobile applications Composites: Part A 42 (2011) 1694–1709

[1-8] A. Naskar, D. Warren
Lower Cost Carbon Fiber Precursor
Oral Presentation - Oak Ridge National Lab ,
Oak Ridge, TN, USA, 16.05.2012

[1-9] F. Baker
Low Cost Carbon Fiber from Renewable Resources
Oral presentation - U.S. Department of Energy, Washington 09.06.2010

2 Production Process of Carbon Fiber

The fundamental component of every carbon fiber is the precursor. The ideal precursor material for making a carbon fiber has a high carbon content, is easy to spin to a precursor fiber, and convertible to a carbon fiber.

The precursor must also be as economic and ecological as possible in producing carbon fiber. There are several promising precursors. Most of the current research is done with a polyethylene, cellulose, and lignin precursors. None of these precursors are commercialized at this moment.

Researchers continue to look for carbon fiber precursors with low cost, high carbon content and if possible, a renewable resource. The major contributors to the high cost of carbon fibers are the precursors, the energy used and the capital equipment for the conversion of the precursor.

Since the early work of Thomas Edison and his use of cellulosic based carbon fiber filaments in the light bulb, many types of precursors have been proven to produce carbon fiber [2-1]. There are two commercial available precursors: Polyacrylonitrile (PAN) and pitch. The most popular precursor to date is polyacrylonitrile (PAN). PAN-based carbon fibers dominate the market, accounting for more than 90% of all sales worldwide. This is due to the fact that PAN is readily available with consistent quality. Pitch based carbon fiber accounts for nearly 10% of worldwide produced carbon fibers. They are used for special applications like high modulus fibers [2-2].

A commercial polyacrylonitrile (PAN) based carbon fiber is produced by the oxidative stabilization process of the precursor and followed by a two-step carbonization process. For high modulus fiber a third heat treatment stage is added. The aim is to remove all chemical impurities in gaseous form. At the end of the manufacturing process, the fiber consists of more than 95% carbon.
The heat treatment is the conventional process used for converting PAN fiber into carbon fiber. Approximately 90% of all commercial available carbon fibers are produced by thermal conversion of a PAN precursor. There are three major steps to

convert the precursor into a carbon fiber [2-3]:

I. Oxidative stabilization (form the stable ladder structure which is needed for processing at higher temperatures)

II. Carbonization (<1600°C – keeps out non carbon atoms and forms the turbostratic structure)

III. Graphitization (>2000°C for improving the orientation of the fiber)

The production process of the lignin based carbon fiber follows the same production steps as PAN based carbon fiber. The production process of a PAN based carbon fiber is schematically shown in Figure 5 [2-3].

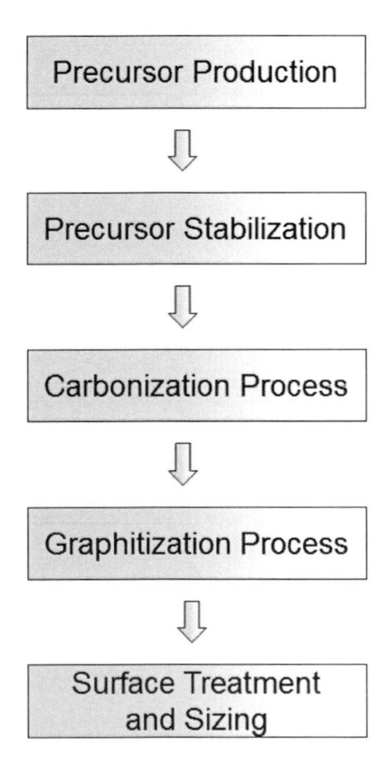

Figure 6: Carbon Fiber Conversion Process [2-3]

2.1 Precursor Production

Before converting a precursor into carbon fiber, it is necessary to produce a precursor fiber. Two major processes exist for producing precursor fibers: the melt spinning and the wet spinning process. Wet spinning is used for most commercial PAN based carbon fibers [2-4]. Since the degradation point of PAN is below the melting point, melt spinning of a plasticized PAN has been used but has never become a commercial process [2-5].

2.1.1 Wet Spinning

East et al. gave an overview concerning the wet spin technology [6]. Figure 6 on page 12 shows the typical wet-spun precursor technology using continuous solution polymerization and wet spinning. It is clear that there are two major steps describing the wet spinning process: dope preparation (polymerization) and the coagulation stage.

Dope preparation (Polymerization): when preparing the dope, acrylonitrile (in powder form) and catalysts are dissolved in solvents like N,N-dimetylformamide (DMF), dimethylsulfoxide (DMSO), sodium thiocyanate (NaSCN), and Zinc chloride ($ZnCl_2$). The concentration of the dope is about 15-20%. To prevent the dope from solidifying, temperatures between 25 and 120°C are required. The dope is pumped into a separate container called the spin bath [2-7].

Coagulation stage: Spinnerets for wet spinning have between 50,000 and 500,000 holes with diameters between 0.05 and 0.25mm. The temperature of the spinning bath is between 0 and 50°C. The line speed of the coagulation varies from 3 to 16m/min. PAN coagulates in the spin bath and the fibril structure is formed [2-8].

Figure 7 summarizes the wet-spun precursor technology using continuous solution polymerization (dope-preparation) and wet spinning (coagulation) [2-6]. (The major process chain follows the red arrows.)

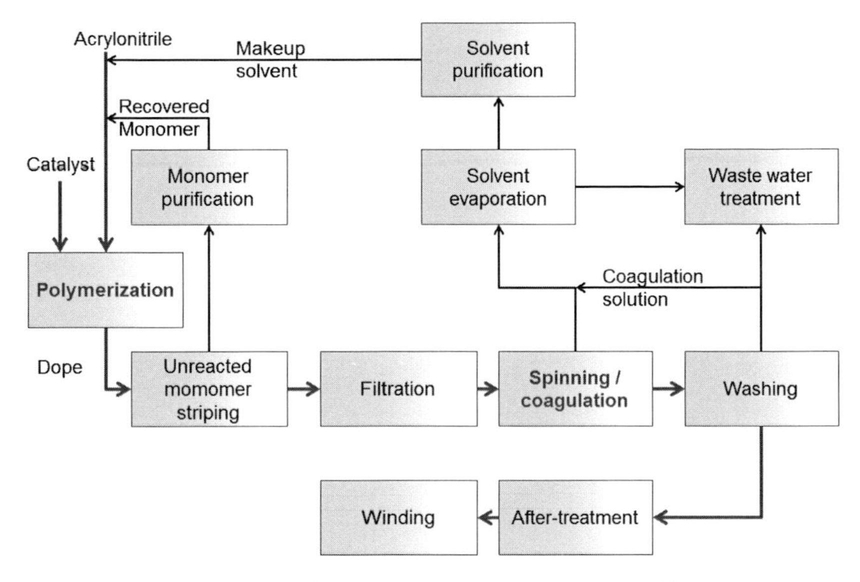

Figure 7: Wet spun precursor technology [2-6]

2.1.2 Melt Spinning

The conventional melt spinning process is useable for materials like lignin which has a melting point temperature below the degradation point. The decomposing of PAN starts below the melting point, which is at about 350°C. Due to PAN's degradation temperature being lower than its melting point, it is impossible to use the conventional melt spinning processes. To perform the melt spinning process a new procedure had to be developed. It was shown in laboratory scale that a homogeneous single phase fusion melt containing water and PAN would allowed the melt spinning process of PAN precursor. This melt can be extruded into a steam pressured solidification zone, which prevents the degradation of the PAN precursor. The melting point of this homogeneous single phase solution drops down to 135°C to 185°C in the steam pressured solidification zone. The extrusion is successfully made with a pressure between 30 and 70 bar using a spinneret with 60-160 µm hole size.

Although the melt spinning process can be used to produce PAN fibers, these fibers contain more internal and external defects when compared to the wet spun PAN fibers. This is the main reason why wet spun PAN fibers used to produce carbon fiber is more common [2-9].

Figure 8 provides an overview of the melt spun technology process as described above [2-6]. (The major process chain follows the red arrows.)

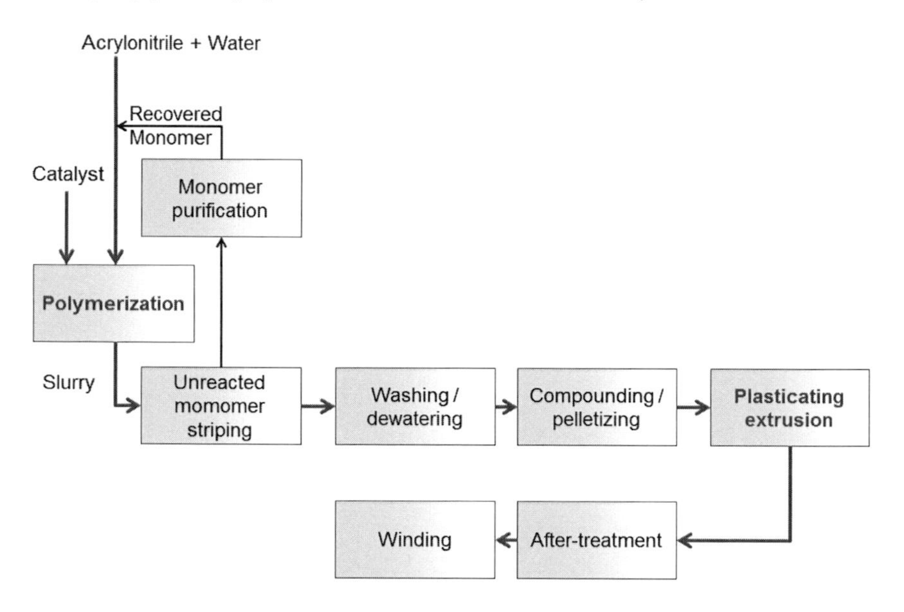

Figure 8: Melt spun precursor technology [2-6]

The comparison of the wet spun precursor technology (Figure 7) and the melt spun precursor technology (Figure 8) shows the higher complexity of the wet spun process compared to the melt spinning process. It should be noted that steps like waste water treatment, solvent evaporation, and solvent purification in the wet spinning process are energy intensive steps and lead to the higher cost of the wet spinning process. Due to the high cost of the wet spinning process, the most common procedure in producing a low cost carbon fiber is using the melt spin technology.

2.1.3 Chemical Structure of a PAN Fiber

Polyacrylonitrile (PAN) is a synthetic, semi crystalline organic polymer resin, with the linear formula $(C_3H_3N)_n$. The chemical structure of PAN in general is planar and is shown in Figure 9 [2-11]. Polyacrylonitrile is a thermoplastic, which does not melt under normal conditions. It degrades before melting. It melts with temperatures above 300 °C if the heating rates are 50 degrees per minute or above [2-10]. Almost all polyacrylonitrile resins are copolymers made from mixtures of monomers with acrylonitrile as the main component.

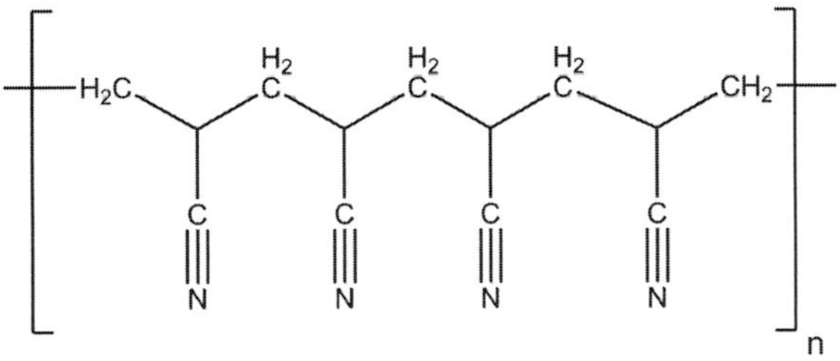

Figure 9: Chemical structure of PAN [2-11]

The cyanide group (-CN) has a large dipole moment. Cyanide is any chemical compound that contains monovalent combining group CN. This group consist of a carbon atom triple-bonded to a nitrogen atom. The electronegativity of the elements of carbon (Pauling scale: 2.55) and nitrogen (Pauling scale 3.04) are very different and leads to a partial positive charging of the carbon and a partial negative charging of the nitrogen atom in this bonding (Figure 10) [2-12].

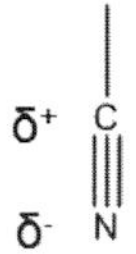

Figure 10: Dipole moment of the cyanide group [2-12]

The opposite electrical charges of the carbon and the nitrogen atoms pull at each other, resulting in an electromagnetic interaction between the cyanide groups. This effect leads to the chemical structure of a PAN Fiber (Figure 11) [2-13].

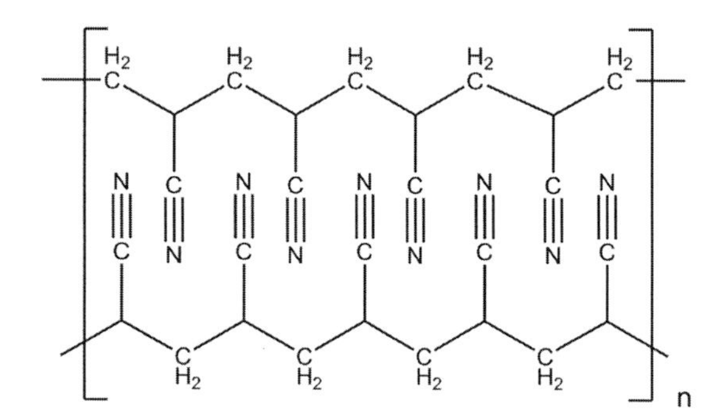

Figure 11: Chemical structure of a PAN-Fiber [2-13]

2.2 Precursor Stabilization

Stabilization is the fundamental process converting PAN into an infusible, non-flammable fiber. Stabilization is performed by heating the precursor fiber at 200-300°C in an oxygen atmosphere. During stabilization the white fiber turns golden yellow, then orange, and then tan-brown and finally black. The stabilization process is an exothermic process, for that reason an elaborate heating program is used to prevent the fibers from burning. It is important to ensure that a sudden release of heat does not occur. This could result in excessive weight loss and chain scission or even fusion [2-14].

The oxidative stabilization of the PAN precursor can be explained by chemical reactions like cyclization, dehydrogenation, aromatization, oxidation, and crosslinking. This leads to the resultant of the stabilized fiber in a ladder structure. The modern understanding of this process is based on the belief that the nitrile groups undergo a cyclization reaction.

Cyclization is the most important step in PAN fiber stabilization. It is the reaction of the nitrile group in the precursor to form the stable ladder polymer of the stabilized fiber. Figure 12 shows the cyclization reaction, which convert the triple bond structure of the CN-group to a double bond structure resulting in a six-member heterocyclic pyridine ring [2-15].

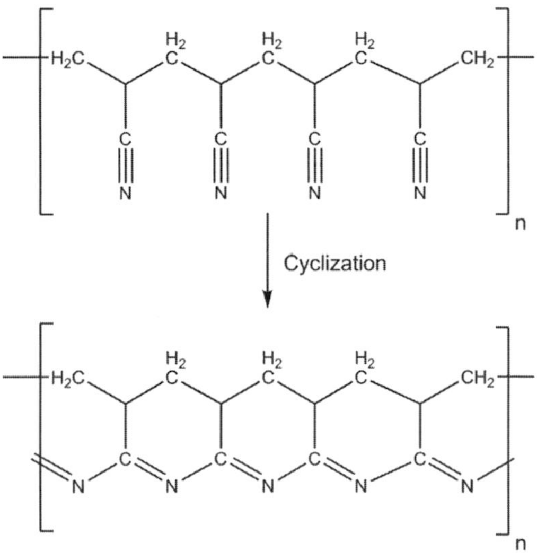

Figure 12: Cyclization of PAN during stabilization process [2-15]

The dehydrogenation process during the stabilization of a PAN fiber contains two steps: the oxidation step and the elimination of water. Dehydrogenation is the process which formats double bonds that stabilize the carbon chain of the PAN fiber. Figure 13 shows schematically the dehydrogenation reaction [2-16].

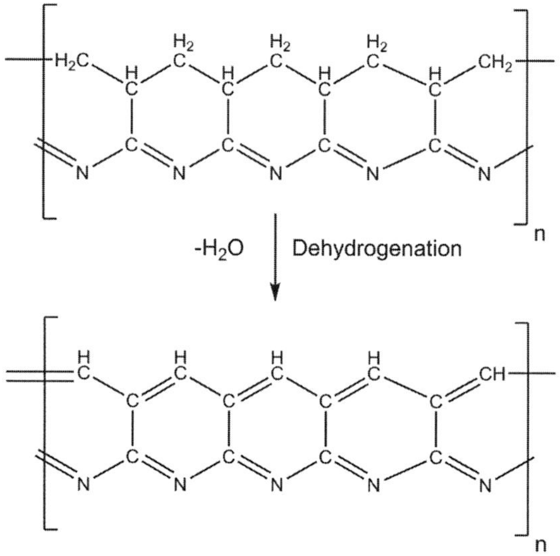

$-H_2O$ Dehydrogenation

Figure 13: Dehydrogenation of PAN during stabilization process [2-16]

The oxidation reaction during PAN fiber stabilization is one of the most complicated reactions. The oxidation of the PAN fiber, which is already undergone the cyclization reaction, is a four step process. The reaction starts with an oxidation creating a hydoperoxide followed by dehydrogenation of water by building a keto-formation. A two-step tautomeric change is responsible for a mixture of hydroxy pyridine and pyridone structures in the PAN ladder polymer. The result is a polymer backbone which contains oxygen-bearing groups that stabilizes the PAN ladder structure and provides greater stability for the high temperature treatment in the carbonization step.

The reaction mechanism is given in Figure 15 [2-17].

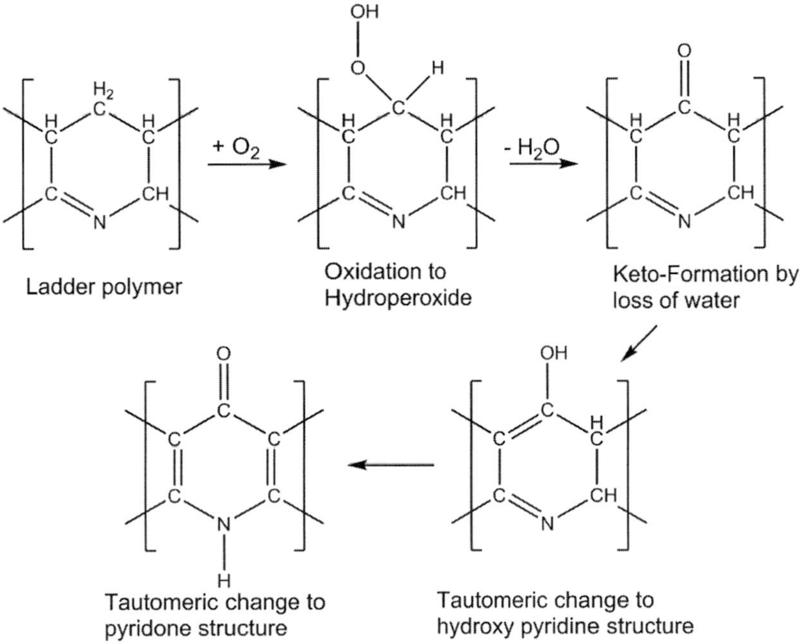

$+ O_2$ → $- H_2O$ →

Ladder polymer

Oxidation to Hydroperoxide

Keto-Formation by loss of water

Tautomeric change to pyridone structure

Tautomeric change to hydroxy pyridine structure

Figure 14: Oxidation of PAN during stabilization process [2-18]

Bansal et al. summarized the stabilization reactions of PAN fiber similar to Figure 14. It becomes clear, that the cyclization and dehydrogenation are processes which run in parallel and at the same time. The oxidation reaction leads finally to a mixture of four different main structures in the stabilized PAN fiber. These main structures are pyridone (40%), pyridine (20%), piperidine (20%), and keto–enol tautomerism forms of pyridine and piperidine (10%) [2-18, 2-19].

The oxidative stabilization is one of the most complicated stages, since different chemical reactions take place at the same time and the structure of carbon fiber is set in this stage.

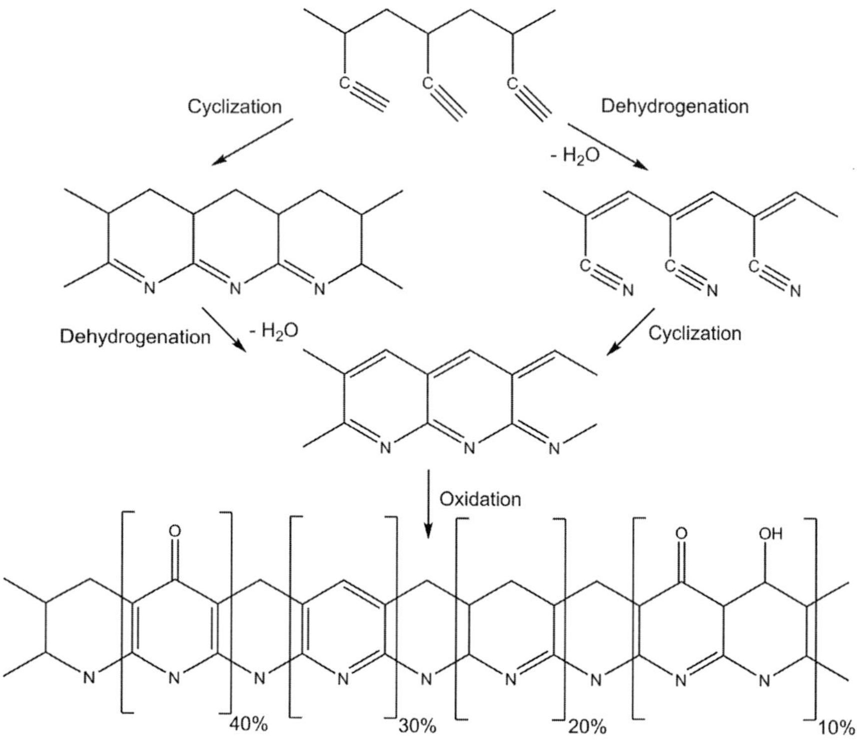

Figure 15: Proposed chemistry of PAN stabilization [2-17, 2-19]

2.3 Carbonization Process

Carbonization is typically done at temperatures of up to 1500°C. Carbonization at 1000°C creates low modulus fibers, while temperatures at 1500°C create intermediate modulus fibers. The process must occur in an inert atmospheric condition, such as a nitrogen rich environment. Argon is commonly used to produce high modulus fibers. The carbonization can be divided into two steps: thermal dehydrogenation at temperatures of up to 600°C and thermal denitrogenation at temperatures of up to 1500°C. In the first step a heating rate lower than 5°C/min is used for dehydrogenation. The second step uses higher heating rates to reach the final high temperatures. The denitrogenation process takes normally 10 min. During pyrolysis the fiber is stretched to increase the modulus and improves the fiber strength, by alining the orientation of the graphitic micro structure. Both steps of the carbonization process are shown in Figure 16 (Dehydrogenation) and Figure 17 (Denitrogenation) [2-18].

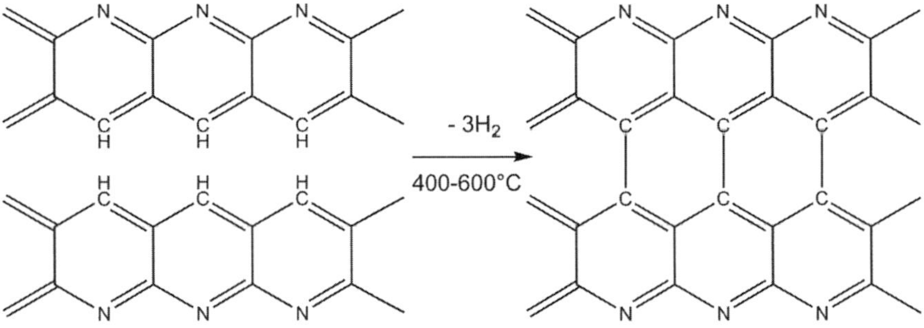

Figure 16: Carbonization step 1: Dehydrogenation [2-18]

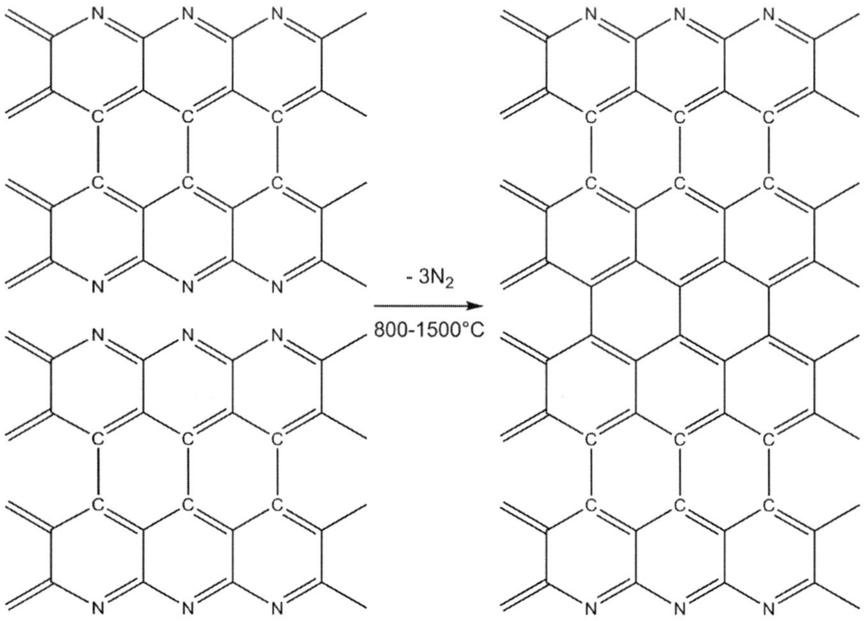

Figure 17: Carbonization step 2: Denitrogenation[2-18]

2.4 Graphitization Process

To produce a higher modulus carbon fiber, it is possible to add a graphitization step. Graphitization is the transformation of the carbon structure into a graphite structure by an additional high temperature treatment using temperatures of up to 3000°C. Similar to the carbonization process, an inert atmosphere is necessary. Generally nitrogen is the preferred inert atmosphere but argon can be used. Argon is eight times more expensive than nitrogen but it creates a carbon fiber with improved strength. At this temperature more than 99% of the PAN is converted into carbon. Carbon fibers which were produced in these conditions are fibers with a very high modulus [2-14].

The structural changes during the carbonization and graphitization process of a PAN fiber to a carbon fiber are shown in Figure 18 [2-23].

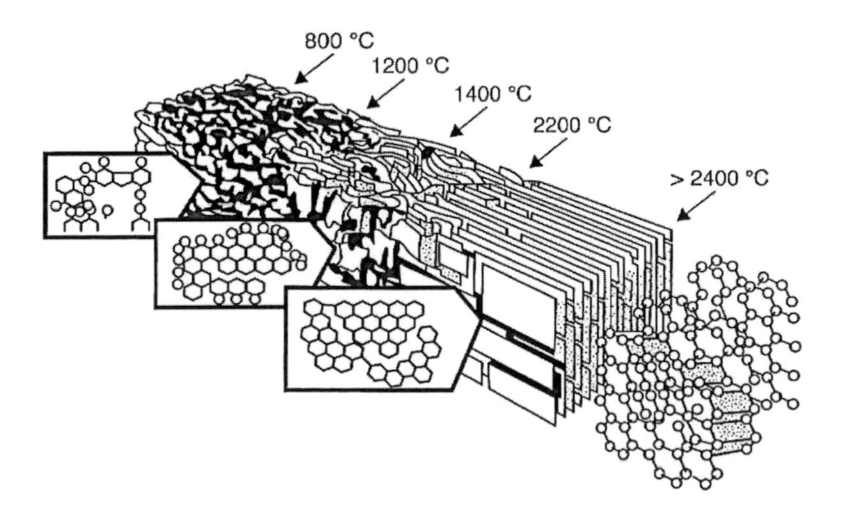

Figure 18: Structural changes of a PAN fiber during heat treatment

2.5 Surface treatment and Sizing

Carbon fibers are mostly used in composites. Stress is transferred inside the composite from one filament of the carbon fiber to the other by the matrix material. Therefor good fiber resin bonding is necessary. If this bonding is weak, the results are poor properties such as low interlaminar shear strength. Generally the problem of poor fiber matrix bonding can be overcome by surface treatment and sizing. There are many different practices of oxidation treatments used such as gaseous, solution, electrochemical, plasma, and catalytic. Non-oxidative surface treatment include the depositing of an active form of carbon, the depositing of paralytic carbon, and grafting by a polymer onto the fiber surface [2-20].

The effects of surface treatment can be summarized as followed: There is possibly little change in the surface area of the carbon fiber. In most cases a weak surface layer is removed and the chemical modification of the surface is detectable by the following groups: -OH, =O, =C=O, -COOH, $-CO_3$. These groups at the surface are responsible for an increase of the polarity of the surface as well as an increase of free energy of the surface of the carbon fiber. A carbon fiber with these active groups on the surface is able to create chemical bonding to the resin which results in higher properties for the composite as described above [2-21].

The application of a coating on the carbon fiber is called sizing and can be achieved by precipitation from solution of a polymer, by precipitation of a polymer onto the carbon fiber surface by electrodepositing, by precipitation of a polymer onto the carbon fiber surface by electro polymerization, or by plasma polymerization. Sizing improves inter-filamentary adhesion and can act like a lubricant preventing fiber damages during textile processing [2-22].

2.6 References

[2-1] T. Edison; US 470925 1880-03-15

[2-2] A. Gupta, I. R. Harrison: New aspects in the oxidative stabilization of PAN-based carbon fibers. Carbon 1996; 34: 1427-1445

[2-3] J. Lui, P.H. Wang, R.Y. Li; Continuous carbonization of polyacrylonitrile based oxidized fibers: aspects on mechanical properties and morphological structures. Appl Polym Sci 1994; 52: 945-950

[2-4] J. Chen, C. Wang, X. Dong, H. Liu: Study on the Coagulation Mechanism of Wet-Spinning PAN Fibers. Journal of Polymer Research 2006; 13: 151-519

[2-5] R. Moreton, W. Watt: The spinning of polyacrylonitrile fibres in clean room conditions for the production of carbon fibres. Carbon 1974; 12: 543-554

[2-6] G.East, J. Mcintyre, G. Patel: The dry-jet wet spinning of an acrylic fiber yarn. J. Text Inst. 1984, 75: 196-200

[2-7] D. R. Paul; A study of spin ability in the wet-spinning of acrylic fibers. Appl Polym Sci 1968; 12: 2273-2298

[2-8] D. R. Paul; Diffusion during the coagulation step of wet-spinning. Appl Polym Sci 1968; 12: 383-402

[2-9] H. C. Hyang, K. Bo-Hye, H. L. Sung, Y. S. Kap: Preparation of Carbon Fiber from Melt Spinnable PAN Co-polymer. Journal of the Korean Chemical Society 2013; 57: 289-294

[2-10] A. K. Gupta, D. K. Paliwal, P. Bajaj: Melting behavior of acrylonitrile polymers. Journal of Applied Polymer Science 1998; 70: 2703-2709

[2-11] K. Morita, Y. Murata, A. Ishitani, K. Murayama, T. Ona, A. Nakajima: Characterisation of commercial available PAN (Polyacrylonitrile)-based carbon fiber. Pure & Appl. Chem. 1986; 58: 455-468

[2-12] K. Kunimatsu, H. Seki, W.G. Golden: Polarization-modulated FT IR spectra of cyanide adsorbed on a silver electrode. Chemical Physics Letters 1984; 108: 195-199

[2-13] G.H. Olive, S. Olive: Intermolecular nitrile reaction between paired dipoles. Polymer Bulletin 1981; 5: 457-469

[2-14] .J. Sánchez-Sotoa, M.A. Avilésa, J.C. del Rio, J.M. Ginésc, J. Pascuald, J.L. Pérez-Rodrígueza: Thermal study of the effect of several solvents on polymerization of acrylonitrile and their subsequent pyrolysis. J Anal Appl Pyrolysis 2001; 58: 155-172

[2-15] R.C. Houtz: "Orlon" Acrylic Fiber: Chemistry and Properties. Textile Research Journal November 1950; 20: 786-801

[2-16] E. Fitzer, D.J. Müller: The influence of oxygen on the chemical reactions during stabilization of PAN as carbon fiber precursor. Carbon 1975; 1: 63-69

[2-17] W. Watt, W. Johnson: Mechanism of oxidization of polyacrylonitrile fibers. Nature 1975; 257: 210 - 212

[2-18] E. Fitzer, W. Frohs, M. Heine: Optimization of stabilization and carbonization treatment of PAN fibers and structural characterization of the resulting carbon fibres. Carbon 1986; 4: 387-395

[2-19] R. C. Bansal, J. B. Donnet: Carbon Fibers 2nd Ed.; Marcel Dekker Inc.: New York, 1994.

[2-20] T. D. Lawrence, J. Michael, R. Lloyd, P. F. Lloyd: Adhesion of Graphite Fibers to Epoxy Matrices: I. The Role of Fiber Surface Treatment. The Journal of Adhesion 1983; 16: 1-30

[2-21] J. D. H. Hughes: The carbon fiber/epoxy interface—A review. Composites Science and Technology 1991; 41: 13-45

[2-22] L.-G. Tang, J. L. Kardos: A review of methods for improving the interfacial adhesion between carbon fiber and polymer matrix. Polymer Composites 1997; 18: 100-113

[2-23] H. Jäger, T Hauke: Carbonfasern und Ihre Verbundwerkstoffe: Herstellungsprozesse, Anwendungen und Marktentwicklung. Süddeutscher Verlag onpact; Stuttgart 2010)

3 Precursor for Carbon Fiber

3.1 Commercial available Carbon Fiber Precursors

Since the early work of Thomas Edison and his use of cellulosic based carbon fiber filaments in the light bulb, many types of precursors have been proven to produce carbon fiber [3-1]. There are two commercial available precursors: Polyacrylonitrile (PAN) and pitch. The most popular precursor to date is polyacrylonitrile (PAN). PAN-based carbon fibers dominate the market, accounting for more than 90% of all sales worldwide. This is due to the fact that PAN is readily available with consistent quality. Pitch based carbon fiber accounts for nearly 10% of worldwide produced carbon fibers. They are used for special applications like high modulus fibers [3-2].

3.1.1 Polyacrylonitrile

Acrylonitrile is made by the Sohio-Process, which is a catalytic reaction of propylene, ammonia and oxygen. The production process of a PAN-based carbon fiber is described in Chapter 2. In general there are three major properties of PAN which makes it an ideal carbon fiber precursor. PAN is a polymer with a continuous carbon backbone and the structure is optimal for cyclization. The carbon content of acrylonitrile (CH_2=CHCN) is 67.9% and during the process of going from a precursor to carbon fiber, about 50% of its mass is lost. This produces carbon fiber with excellent properties [3-3]. The chemical structure of PAN with a continuous carbon backbone is shown in Figure 19 [3-4].

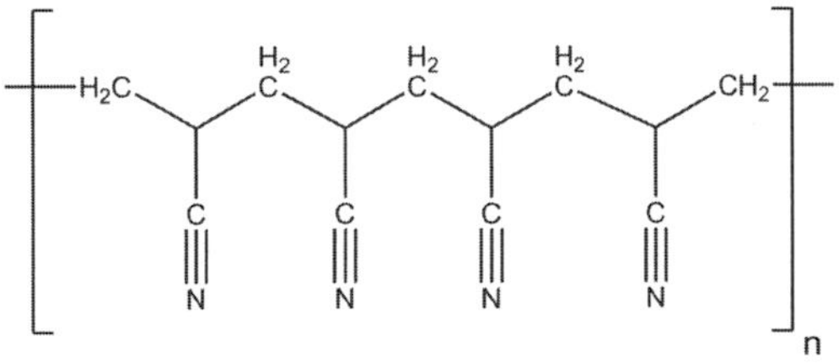

$$\left[-H_2C-CH(C{\equiv}N)-CH_2-CH(C{\equiv}N)-CH_2-CH(C{\equiv}N)-CH_2-CH(C{\equiv}N)-CH_2- \right]_n$$

Figure 19: Chemical structure of PAN

The two main components of acrylonitrile are propylene and ammonia. It is generally recognized that the price of propylene is directly linked to the crude oil price, which dictates the price of PAN. The chemicals for the production of PAN are approximately 45% of the cost of making the precursor. The high price of the PAN precursor and the big carbon footprint combined with a conversion rate of 50% are the main disadvantages for making PAN based carbon fiber [3-5].

3.1.2 Pitch

Pitch is a tarry material which can be produced from many different sources and is solid at room temperature. There are five main sources for pitch: petroleum refining, destructive distillation of coal, natural asphalt, pyrolysis of polyvinylchloride (PVC) and pyrolysis of naphthalene or anthracene.

Pitch is a complex mixture of hundreds of aromatic hydrocarbons with a molecular mass ranging from 300 to 1200. In general pitch is formed out of four main classes of chemical compounds: saturates (low molecular weight aliphatic compounds), naphthenes (low molecular weight aromatics), polar aromatics (higher molecular weight), and asphaltenes (highest molecular weight and thermally most stable). A high concentration of asphaltenes in the petroleum pitch works better for conversion to carbon fiber. The carbon content of pitch is an average of 85%. This process produces carbon fiber with a mass loss of 60-75% which makes it suitable for manufacturing high modulus carbon fiber. The conventional process of producing a pitch based carbon fiber is illustrated in Figure 21, and an example for a typical structure of a pitch precursor is shown in Figure 20 [3-6, 3-7].

When comparing pitch based carbon fiber to PAN based carbon fiber, it is important to consider price, mass loss, and carbon foot print. Pitch and PAN based precusors are both derived from crude oil. As the price of crude oil continued to rise in the future, so will the price of pitch and PAN precursors. With the mass loss of 60-75% during conversion of pitch to carbon fiber, the production cost of this fiber is more expensive. Pitch based carbon fiber has also a high carbon footprint related to the mass loss during conversion and the precursor preparation methods which uses a high amount of energy.

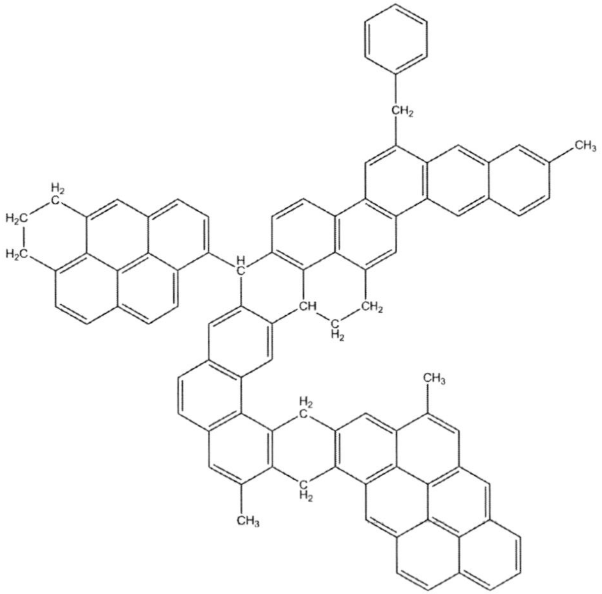

Figure 20: Chemical structure of a pitch precursor [3-6]

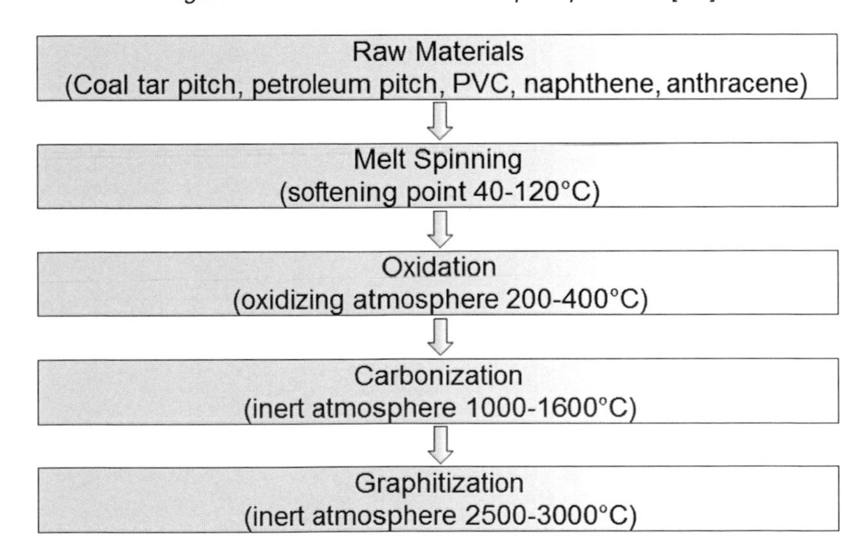

Figure 21: Conventional carbon fiber production process using a pitch precursor [3-7]

3.2 Alternative Precursors

The ideal precursor material for making a carbon fiber should have a high carbon content, is easy to spin to a precursor fiber, and convertible to a carbon fiber. The precursor must also be as economic and ecological as possible in producing carbon fiber. There are several promising precursors. Most of the current research is done with a polyethylene, cellulose, and lignin precursors. None of these precursors are commercialized at this moment.

3.2.1 Polyethylene

Polyethylene (PE) is the most commonly used plastic today. The annual global production is approximately 80 million tons [3-8]. There are many different kinds of polyethylene plastics, with most having the chemical formula $(C_2H_4)_nH_2$ and a carbon content of 86%. Thus PE is usually a mixture of similar organic compounds that differ in terms of the value of n. The general chemical formula of polyethylene is given in Figure 22.

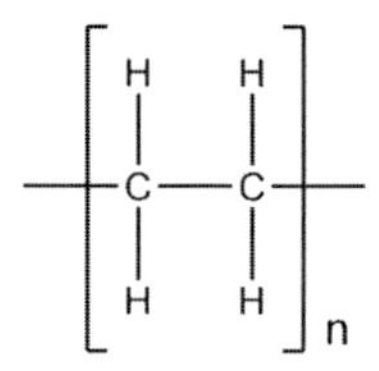

Figure 22: Chemical structure of Polyethylene [3-8]

Polyethylene can not be stabilized by the conventional oxidation process. An oxidation reaction of polyethylene will decompose the precursor fiber and produce the by-products; water (H_2O) and carbon dioxide (CO_2). For this reason the invention of new stabilization processes is necessary. There are two ways to stabilize polyethylene: by treating the precursor fiber with chlorosulfuric acid (HSO_3Cl) at 60-90°C or by treating it with 98% Sulfuric acid (H_2SO_4) at 100-180°C. These treatments create a higher temperature resistance of the polyethylene precursor fiber and make the carbonization step possible [3-9].

The mass loss during the carbonization of polyethylene precursor fiber, is between 20 and 50%. An important paper written by Zang et al. shows the investigation of the structure and the development of the properties during conversion of PE in a laboratory process [3-10].

Similar to pitch and PAN based carbon fiber precursors; the price of conventional polyethylene is directly linked to the crude oil price which will rise in the near future. There are possibilities for producing polyethylene from renewable resources to create an independence from the crude oil price and lower the carbon footprint of the polyethylene precursor. The mass loss of 20-50% during conversion of polyethylene to a carbon fiber makes this precursor an attractive alternative. Since polyethylene has a high carbon content of 86%, polyethylene has a higher yield percentage than using a conventional precursor.

The biggest problem in the commercialization of polyethylene as a carbon fiber precursor will be the enormous amounts of waste acid. This waste acid, from the stabilization process, needs to be recycled or will need to be recovered. Until this problem is solved or an alternative stabilization process is found, the commercialization of polyethylene will be difficult.

3.2.2 Cellulose

Cellulose is the oldest available precursor for carbon fiber since 1880. Its use was first documented by Thomas Edison during his development of electric lamp filaments [3-1]. Since the 1950's, man-made fibers containing regenerated cellulose has been referred to as rayon. Cellulose has the chemical formula $C_6H_{10}O_5$ with the chemical formula given in Figure 23.

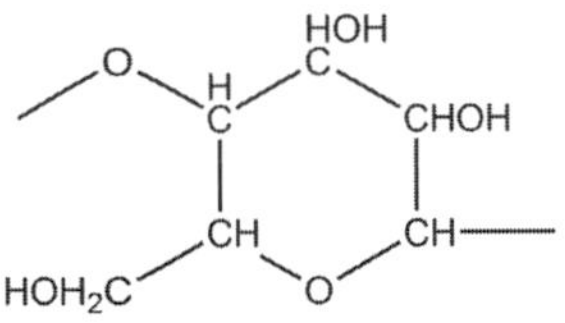

Figure 23: Chemical structure of cellulose

The specific processes and data for making carbon fiber from cellulose and all other precursors is confidential information each producer withholds. For cellulose based carbon fiber only some major steps from the viscose rayon process are public. The viscose rayon process contains the following three major steps: stepping, xanthation and spinning. An outline of a typical viscose process is shown in Figure 23 [3-11].

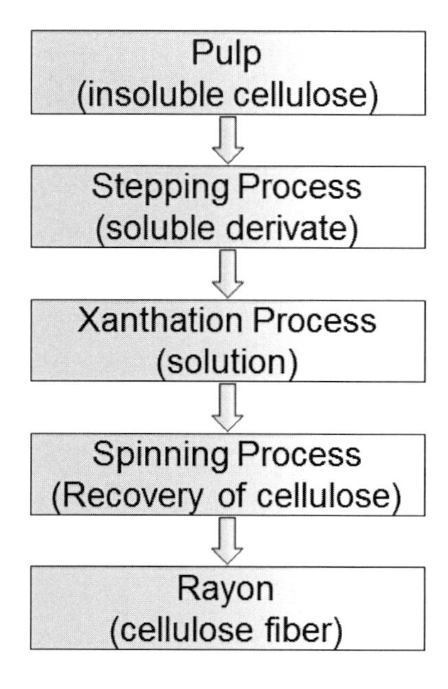

Figure 24: Viscose rayon process

In the first step of the viscose rayon process, which is referred to as the stepping process, the insoluble cellulose is carefully treated under dehydration with sodium hydroxide (NaOH) to make it soluble (Figure 25). The result is alkali-cellulose [3-12].

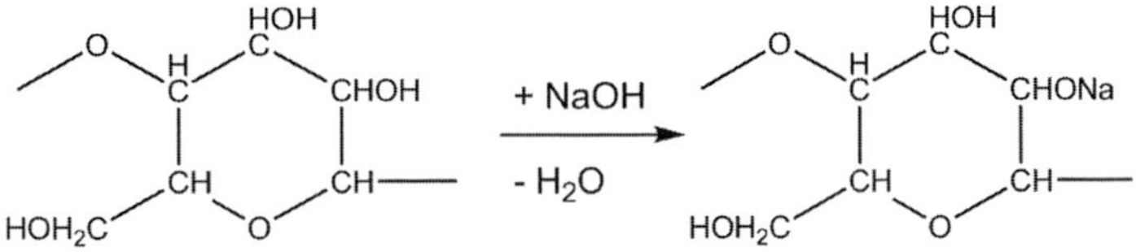

Figure 25: Stepping process

In the xanthation step, carbon disulfide (CS_2) is used like a catalyst to activate the alkali-cellulose for the fiber spinning process. The result is sodium cellulose xanthate in a form of a thick viscous solution (Figure 26) [3-13].

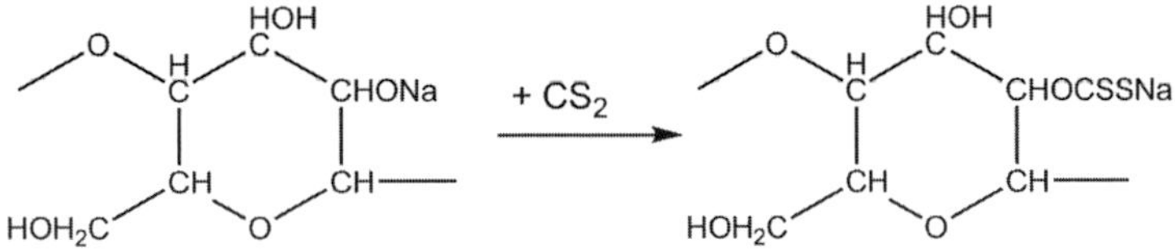

Figure 26: Xanthation process

The higher reactivity of the sodium cellulose xanthate is important for the fiber spinning process. Two units of sodium cellulose xanthate are then combined to the first unit of the rayon fiber. With the help of sulfuric acid (H_2SO_4) the carbon disulfide is then recaptured. The result of this step is the first rayon unit (Figure 27) [3-14].

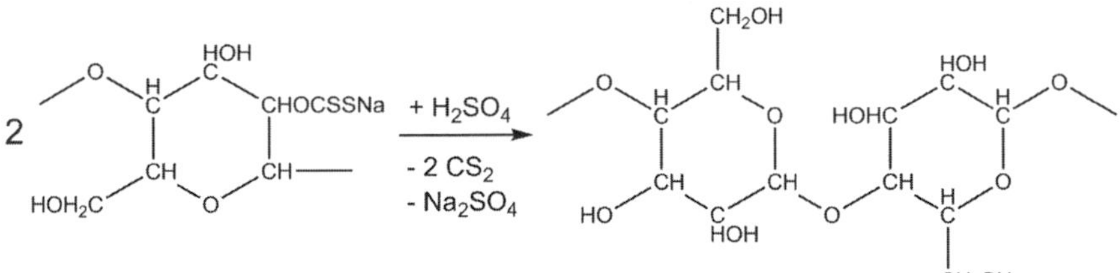

Figure 27: Rayon spinning process

The final product is Rayon. It is a natural polymer containing glucose monomers which are linked by an ether bond. The ether bond is giving a relatively stiff polymer with up to 1000 glucose monomer units and the number of structural fragments (n) equals up to 500. This is shown in Figure 28. With this rayon fiber the carbonization is made [3-15, 3-16, 3-17, 3-18].

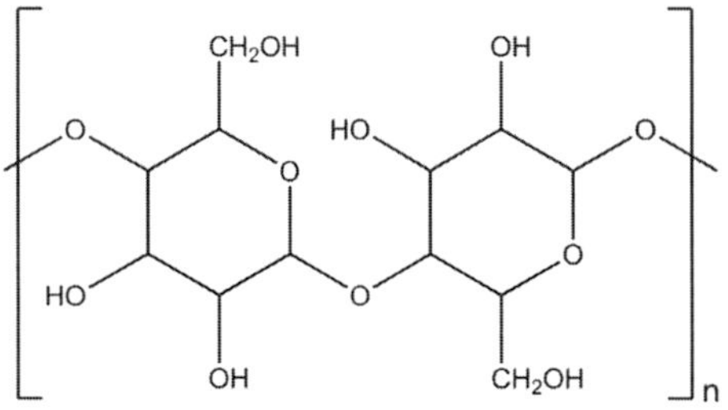

Figure 28: Chemical structure of Rayon made from cellulose

The worldwide cellulose production in 2012 was 13.7 million tons with a price of $1344 per ton [3-19]. That makes cellulose plenty available with a relativity low price compared to the conventional precursors like PAN or pitch. Also as a renewable resource the carbon footprint is lower than the footprint of commercial precursors.

The biggest disadvantage of cellulose as a precursor is the low carbon content of 45% compared to other precursor (PAN 68%; pitch 85%). A low conversion rate of around 20-30% is the consequence of the low carbon content. The combination of the low carbon content, the low conversion rate, the higher cost per ton, and poor fiber properties make cellulose less attractive. To commercialize cellulose precursors, improvements in their mechanical properties and the fiber spinning process need to be made.

3.2.3 Lignin

The use of the raw material, lignin, as a precursor for carbon fiber is still relatively unknown. Lignin combines some of the best economical and ecological properties of any precursor material currently available. The cost benefit and the renewability, as well as the high availability of lignin are promising characteristics. This is why today's research activities within the Volkswagen Group are focusing on this material.
Lignin is a waste product from the paper mill industry. The price of lignin is equal to the value of the energy which is produced by burning it (fuel value: 23.4 MW/kg). The worldwide production in 2013 was more than 120 million tons with a price of less than $250 per ton. Comparing lignin with PAN ($2400 per ton) or cellulose ($1344 per ton) cost savings of the precursor of more than 80% is possible. This means that lignin is the cheapest available precursor for carbon fiber currently and there is readily available [3-20].

Since lignin has higher carbon content then cellulose (45%) the conversion rate is also higher (50%). That means lignin has a higher yield than cellulose. When comparing lignin to PAN, pitch and PE the carbon content of lignin is lower, with 67.9%, 85% and 86% carbon content respectively. However, because lignin cost $250 per ton, the dollar to carbon content ration ($0.50 per 1kg carbon) is almost 3 times better then the next cheapest alternative (PE, $1.40 per 1 kg carbon). PAN has a dollar to carbon content ratio of $3.50 per 1 kg carbon.

Also a low carbon footprint makes lignin very attractive as a carbon fiber precursor. Since PAN, PE and Pitch are made from crude oil, the production of a precursor from these sources requires many extraction processes and chemical reactions, which causes a high carbon footprint. As mentioned above, lignin is a waste product of the paper mill industry and is of great abundance and renewable.

Chapter 3.5 gives an overview concerning lignin production and Chapter 4 presents the lignin properties of a Hardwood Lignin isolated by Kraft pulping.

The following chapters will show the development (Chapter 5), properties (Chapter 7.1) and chemical structure (Chapter 7.2) of lignin based carbon fibers made in a laboratory scale process as part of this thesis.

3.3 Lignin as a Precursor for Carbon Fiber

Lignin is an amorphous, heterogeneous, and complex polymer and is the second most important natural polymer. It is produced as a byproduct of the paper mill industry and is the main organic compound of the black liquor, which is a waste of cellulose production. Since the fuel value of lignin is 23.4 MJ/kg, the primary use of lignin in the paper mill industry is waste incineration with thermal dissipation, to produce the processing heat in the paper mill industry. However, the value is increased when used to produce other products. Examples of such products can be found in Table 2. The worldwide production of lignin is estimated to more than 120 million tons per year [3-25].

Table 2: Potential products with lignin as a feedstock

Products with lignin as a feedstock	Technology status	Market price of the product
process heat and energy	H	0,016 €/kWh
synthesis gas	H	vary
synthesis gas products		
Methane	H	0,20 €/l
dimethyl ether (DME)	H	0,20 €/l
Ethanol	L	0,22 €/l
Hydrocarbons		
Cyclohexane	L	0,45 €/l
Styrol	L	0,18 €/l
Phenol		
Phenol	L – M	0,17 €/l
Substituted phenol	L	0,18 – 0,40 €/l

oxidation product		
Vanillin	H	8,77 €/kg
dimethyl sulfoxide (DMSO)	H	1,53 €/kg
aromatic acids	L	0,69 – 0,77 €/kg
aliphatic acids	L	0,60 – 0,98 €/kg
macromolecular products		
carbon fiber	L – M	goal: 7,50 €/kg
Polyelectrolyte	L – M	4,61 €/kg
Polymere filler material	L – M	1,54 €/kg

H - High (high technology level – good research results)
L - Low (low technology level – many research necessary)
M - Medium (medium technology level – first good research results)

There are two major application fields for lignin. One of the applications is the use as fuel and the other is the use for higher valuable products, e.g. lignin based carbon fiber, lignin based hydrocarbons or lignin based electrolyte. Table 2 gives an overview concerning possible products using lignin [3-26].

Less than 1% of the worldwide lignin production is used for basic chemicals or polymers. The main reason is the high inhomogeneity of lignin which makes it difficult to reach a constant level of high product quality in the area of basic chemicals or polymers. Today only vanillin and some phenolic derivates are commercially made from lignin [3-26].

A highly profitable area for products made from lignin could be the production of precursors for carbon fiber in the future. Since the value of lignin used in this area is extremely high (Table 2).

3.4 Chemical Anatomy of Wood and Lignin

Wood has a three dimensional fiber composite structure with the main components [3-22] cellulose, hemicellulose and lignin (Table 3), as well as some minor constituents like phenols, terpenes and organic acids. Lignin can be found with concentrations from 20 to 30% in wood depending on the kind of wood.

Wood is not a homogeneous material, which makes it difficult to describe it from the chemical and biological point of view.

Normally a differentiation is made between hardwood and softwood. Softwood, sometimes also called pinewood, includes spruce, fir, pine and cedar. Examples for Hardwood or deciduous trees and shrubs are oak, beech, maple and birch [3-21].

Table 3: Lignin content of different kinds of wood

Kind of wood	Cellulose (%)	Hemicellulose (%)	Lignin (%)
Softwood	40-45	25-30	25-30
Hardwood	40-45	30-35	20-25

Lignin has a very complex, heterogeneous and branched out structure (Figure 29). Furthermore the structure of lignin can vary a lot and is connected to the kind of wood (hardwood / softwood).

The structure of lignin, which can be used as a precursor for carbon fiber, is also influenced by the isolation process. Different isolation processes are available that lead to different structures of the isolated lignin. Since lignin can be found in two and three dimensional structures in the cells, and the different isolation processes can break different bonds during pulping, the results are different lignin monomers and side chains [3-32].

The chemical characterization of the monomers and side chains of the used Hardwood Lignin for carbon fiber production is given in Chapter 4.2

In general, lignin is a polymer which is built from three major monomers. These three monomers are coumaryl alcohol, coniferyl alcohol, and sinapyl alcohol (Table 4) [3-24]. How these monomers are combined in the side chains of the Harwood lignin, which is used for making carbon fiber, is further discussed in Chapter 4.3.

Table 4: Monomeres of Lignin

coumaryl alcohol	**coniferyl alcohol**	**sinapyl alcohol**
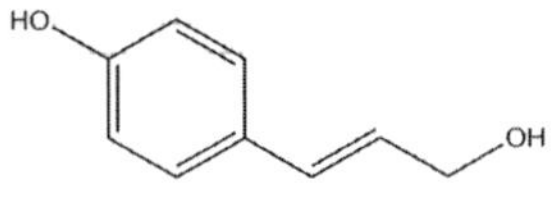 HO, OH	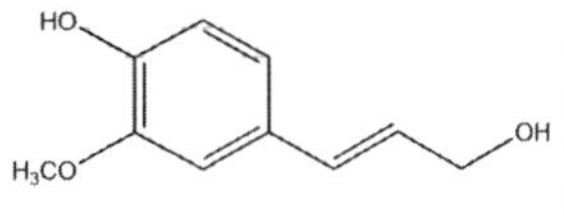 HO, H_3CO, OH	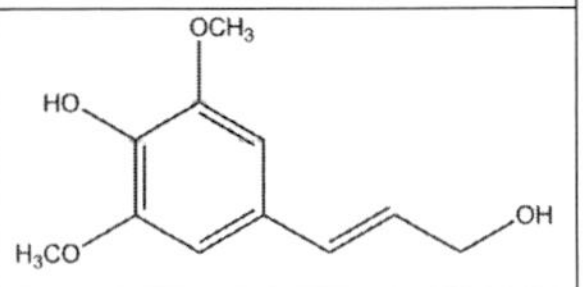 OCH_3, HO, H_3CO, OH

The lignin biosynthesis, which is taking place in every plant, is responsible for the complex structure of the polymer.

During the biosynthesis, the phenyl propane monomer is transferred to a free phenoxy radical by the enzyme peroxidase. This free phenoxy radical is dislocated in the molecule and can be stabilized at the aromatic ring or at the carbon side chain. Because of this dislocation of the radical coupling reactions in form of ether bonds, carbon-carbon bonds, and the bonding of several phenyl propane units take place. This creates the complex structure of the lignin polymer (Figure 29) [3-24].

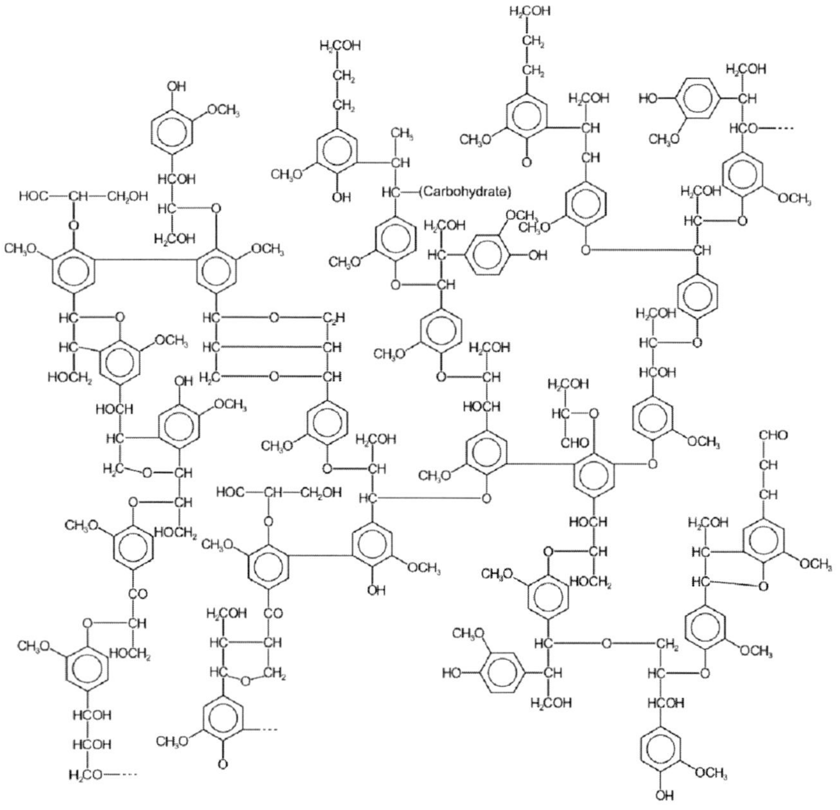

Figure 29: Example of the structure of lignin

3.5 *Isolation of Lignin from Wood*

The focus on more environment friendly materials has prompted the interest in lignin as a precursor. There are five major processes for isolation of lignin from wood: the Kraft Process, the Sulfite Process, the Alcell Process, the Organocell Process and the processes in the Biorefinery.

The pulping industry alone provides on average 128 million tons of lignin as waste every year. The most commonly used process in the pulping industry is the Kraft Process with 75% of the worldwide lignin production. Second most common is the Sulfite process with 8%. The other 17% is produced by special processes (the Alcell Process, the Organocell Process) which are used for the production of special papers and special kinds of cellulose (Table 5) [3-27].

Table 5: Worldwide Lignin Production (in 1000t)

Country	Lignin total	Kraft Process	Sulfite-Process	Other Processes
Germany	1377	0	1023	354
Spain	1506	1375	0	131
France	1727	1349	258	120
Finland	5916	5224	154	538
Sweden	7337	6003	749	585
Rest of EU	4295	2759	588	948
Canada	13195	10850	1627	718
China	8894	1032	0	7862
USA	50775	44106	1418	5251
Other	33295	24265	4067	4963
Worldwide	128317	96963	9884	21470

3.5.1 The Kraft Process

The Kraft pulping process is the most common pulping process in the world, giving large quantities of lignin that is burned in the recovery boiler. More than 75% of the lignin produced worldwide is extracted by Kraft pulping (Table 5 and Figure 30).

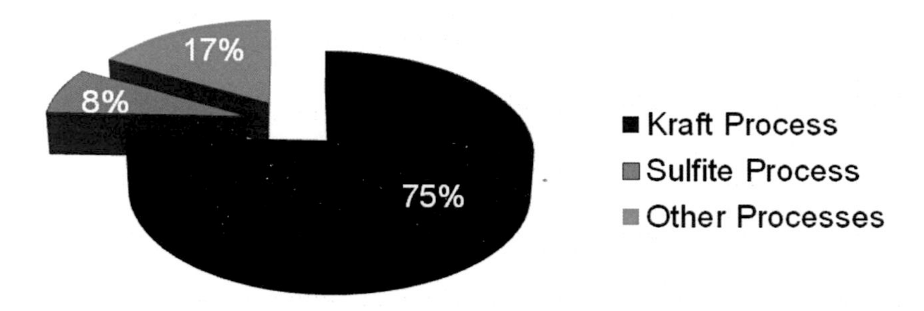

Figure 30: Distribution of worldwide lignin production by process

Partial isolation of the Kraft lignin would be of benefit for the mill, which gains both a new Kraft lignin raw material and the possibility of increasing pulp production [3-28].

The ability to pulp almost every kind of wood is possible with the Kraft Process. That is a big advantage compared to other processes.

First the lignin is treated in an alkaline solution, under high temperatures, which breaks the alkyl-ether-bonds. This creates lignin monomers, which can be found in the form of phenolates (Figure 31). Then the SH^--Ions in the solution are used to prepare the hydrolysis at the α-C-atom at the propane-side-chain and activates the alkyl-ether-bond. During the hydrolysis, some of the methoxy-groups will decompose and form methanol, methyl captan, and dimethyl sulphide [3-29].

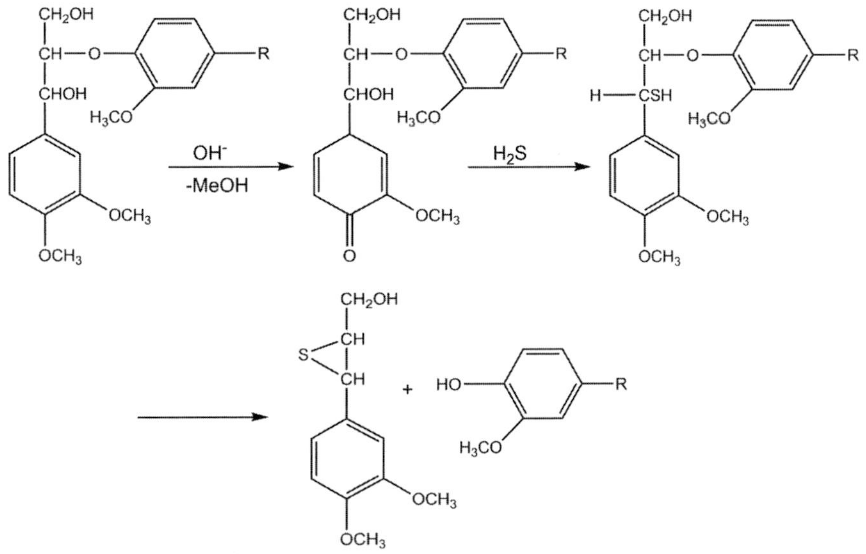

Figure 31: Chemical reactions of the Kraft pulping process (example β-O-4-bonding)

3.5.2 The Sulfite Process

The sulfite process uses a hydrogen sulfite solution for breaking the alkyl-ether-bonds during pulping. With the sulfite process, pulping of wood with low content of natural resin and silicon dioxide is possible [3-28].

Using the sulfite process, lignin is sulfonated and hydrolyzed, resulting with a ligno-sulfonate in a alkaline solution. The phenyl propane groups get sulfonated at the side chain by breaking the bond with the hydroxy and alkyl-ether-bonds (Figure 32). The HSO_3^- -ion breaks the ether bond at the same moment. It is important to regulate the pH-value. If the pH-value is too low the lignin monomers condensates again. If that happens the lignin is not solvable anymore, which is the biggest disadvantage of this process [3-30].

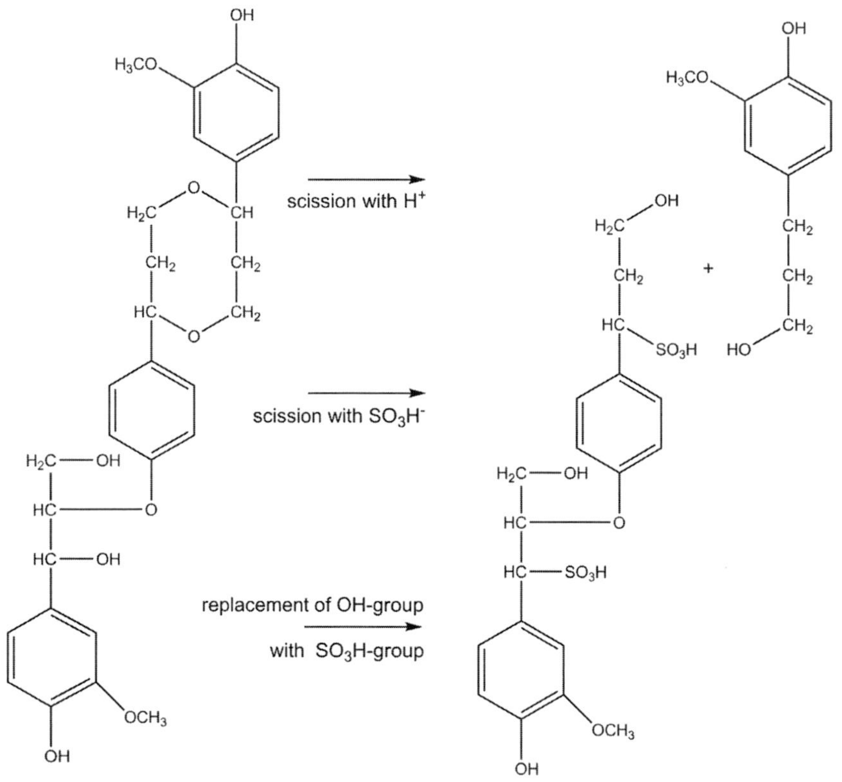

Figure 32: Chemical Reactions of the Sulfite pulping process

3.5.3 The Organocell Process

The Organocell process has been used since 1991 for the production of cellulose and lignin. This process is working by using methanol and sodium hydroxide without using sulfur. The best results achieved during this process are done so using soft-wood. The wood chips are cooked at 200°C in 35 bar pressure using a 1:1 water ethanol solution. After that the solution is treated with 20% sodium hydroxide [3-31].

During the process the ether-bond is broken and the monomeric lignin units are stabilized with methyl groups. As a byproduct the organosolv lignin is produced, which is solvable in many organic solvents. The organosolv lignin has a melting point at approximately 185°C and has an average molar mass of 1000 g/mol [3-32, 3-33, 3-34].

3.5.4 The Alcell Process

The Alcell process is also used for the production of cellulose and organosolv lignin. Like in the organocell process there is no sulfur compound necessary for pulping. This process works best with hardwood. At a temperature of 200°C and a pressure of 35 bar, the wood chips are cooked in a 1:1 ethanol water solution. During the process the ether-bonds are broken and the monomeric lignin units are stabilized with ethyl groups. As a byproduct the organosolv lignin is produced, which is solvable in many organic solvents. The organosolv lignin has a melting point at around 145°C and an average molar mass of 1000 g/mol [3-35, 3-36].

3.5.5 Biorefinery

There is an ongoing change in the pulp and paper industry today due to increasing competition and increasing prices of the wood raw material and oil. Usually, the mills produce only pulp and paper with chemical recycling and internal, and sometimes external, energy production. A biorefinery utilizes the whole raw material by integrating the conversion of biomass with production of biofuel, energy, heat, and bio based products. By utilizing the side streams in the process, these mills have capability to be transformed into biorefineries. In biorefinery the new Ligno Boost process is used for lignin separation which used ultrafiltration and provides very pure lignin [3-37, 3-38, 3-19].

3.5.6 Influence of the pulping process on chemical structure

In order to produce highly valuable lignin based products, the lignin feedstock needs to fulfill some fundamental properties and some defined chemical structural demands. The lignin should have a low price, good quality and good availability. A big advantage would be the consistency in the chemical structure. Also a high number of functional groups are desirable to start chemical reactions with lignin [3-40]. Figure 33 gives an overview concerning quality and availability of lignin made with different processes.

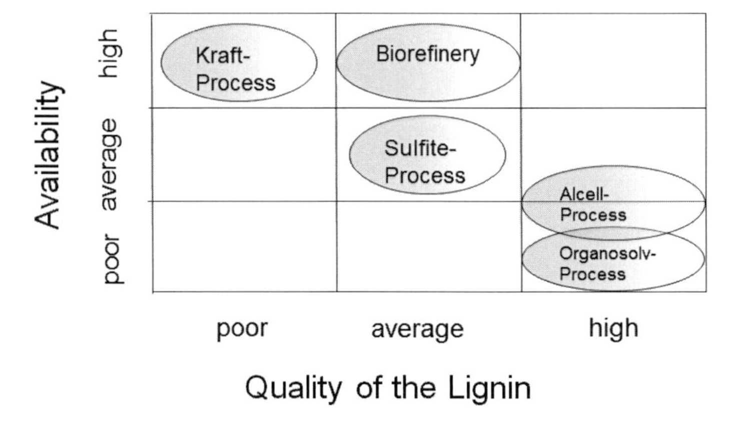

Figure 33: Availability and quality of lignin from different production processes

A disadvantage of the lignin made by the Sulfite Process is that most of the OH-groups are replaced by sulfonic acid groups. The reactivity of Kraft lignin is also reduced by the production process. Compared to sulfite lignin, Kraft lignin has more phenol groups, which means a better reactivity. A major disadvantage for both Kraft and Sulfite lignin, is the sulfur content which can have a negative influence on the reaction. This sulfur creates waste gases containing sulfur during carbon fiber production [3-40]. Lignin produced by the Alcell or Organocell process is sulfur free. During the process of pulping many of the OH-groups get alkylated [3-40]. The reactivity of the lignin can also be influenced by the kind of wood used. For example hardwood has a high content of syringyl alcohol which is blocked for many reactions by ortho substitution with a methoxy group. Softwood Lignin has a higher content of coniferyl alcohol with free phenolic groups in ortho position, which creates higher reactivity. That means hardwood has a lower reactivity than Softwood Lignin in this case [3-40]. To summarize, the choice of the pulping process and the choice of the kind of wood are responsible for the quality of the lignin used for carbon fiber production. Advantages and disadvantages of different pulping processes are summarized in Table 6 [3-40].

Table 6: Advantages and disadvantages of different pulping processes

Process	Advantage	Disadvantage
Kraft-Process	• high availability • low price • high phenol content	• highly sulphurous • contains recondensat • quality fluctuations
Sulfite-Process	• high availability • low price	• highly sulphurous • contains recondensat • low phenol content • quality fluctuations
Alcell-Process	• sulfur-free • constant quality	• low availability • high price • OH-groups blocked by ester bondings
Organosolv-Process	• sulfur-free • constant quality	• almost no availability • very high price • OH-groups blocked by ester bondings
Biorefinery	• high availability • low price • sulfur-free	• quality fluctuations • fluctuation in the raw material base

3.6 References

[3-1] T. Edison; US 470925 1880-03-15

[3-2] A. Gupta, I. R. Harrison: New aspects in the oxidative stabilization of PAN-based carbon fibers. Carbon 1996; 34: 1427-1445

[3-3] Z. Wangxi, L. Jie, W. Gang; Evolution of structure and properties of PAN during there conversion to carbon fibers. Carbon 2003; 41: 2805-2812

[4-4] O.P. Bahl, L.M. Manocha; Characterisation of oxidized PAN fibers. Carbon 1974; 12: 417-423

[3-5] S. C. Chiu, E. J. Ryan: Characterization of fiber structures developed during carbon fiber conversion process. Carbon 2010; 1: 818-819

[3-6] T. Cho, Y. S. Lee, R. Rao, A. M. Rao, D. D. Edie, A. A. Ogale: Structure of carbon fiber obtained from nanotube-reinforced mesophase pitch. Carbon 2003; 41: 1419-1429

[3-7] N. Díez, P. Álvarez, R. Santamaría, C. Blanco, R. Menéndez, M. Granda: Optimisation of the melt-spinning of anthracene oil-based pitch for isotropic carbon fibre preparation. Fuel Processing Technology; 93, 1; 99-104

[3-8] J. Holbery, D. Houston: Natural-fiber-reinforced polymer composites in automotive applications. JOM 2006; 58: 80-86

[3-9] I. Karacan, K. Ş. Tunçel: Thermal stabilization of poly(hexamethylene adipamide) fibers in the presence of ferric chloride prior to carbonization. Polymer Degradation and Stability 2013; 98: 1869-1881

[3-10] D. Zhang, Q. Sun: Structure and properties development during the conversion of polyethylene precursors to carbon fibers. Journal of Applied Polymer Science 1996; 62: 367-373

[3-11] J. Cai, L. Zhang, J. Zhou1, H. Qi, H. Chen, T. Kondo, X. Chen, B. Chu: Multifilament Fibers Based on Dissolution of Cellulose in NaOH/Urea Aqueous Solution: Structure and Properties. Advanced Materials 2007; 19: 821-825

[3-12] T. Okano, A. Sarko: Mercerization of cellulose. II. Alkali–cellulose intermediates and a possible mercerization mechanism. Journal of Applied Polymer Science 1985; 30: 325-332

[3-13] G. A. Stepanova, A. B. Pakshver, A. L. Kaller: Acceleration of the xanthation process of alkali cellulose. Fibre Chemistry 1982; 14: 27-29

[3-14] N. Drisch, P. Herrbach; US2705184 A 1955-06-03

[3-15] M.M Tang, R. Bacon: Carbonization of cellulose fibers—I. Low temperature pyrolysis. Carbon 1964; 2: 211-214

[3-16] M.M Tang, R. Bacon: Carbonization of cellulose fibers. II. Physical property study. Carbon 1964; 3: 390-393

[3-17] M. Sevilla, A.B. Fuertes: The production of carbon materials by hydrothermal carbonization of cellulose. Carbon 2009; 47: 2281-2289

[3-18] Q. Wu, D. Pan: A New Cellulose Based Carbon Fiber from a Lyocell Precursor. Textile Research Journal 2002; 72: 405-410

[3-19] FOOD AND AGRICULTURE ORGANIZATION OF THE UNITED NATIONS FAOStat (2012) Forestat. http://faostat.fao.org/site/626/default.aspx#ancor

[3-20] I. S. Nashawi, A. Malallah, M. A. Bisharah: Forecasting World Crude Oil Production Using Multicyclic Hubbert Model. Energy Fuels; 2010; 24: 1788–1800

[3-21] H. Staudinger, E. Hussemann: Chemische Anatomie des Holzes. Holz als Roh und Werkstoff 1941; 4: 27-29

[3-22] D. N. S. Hon, N. Shiraishi: Wood and Cellulosic Chemistry. ISBN-13: 978-0824700249, November 2000

[3-23] D. Fengel, G. Wegener: Wood; ISBN-13: 978-3110084818 , Juni 2011

[3-24] P. Sitte, E. W. Weiler, J. W. Kadereit, A. Bresinsky, C. Körneror, E. Strasburger: Strasburger - Lehrbuch der Botanik für Hochschulen; ISBN-13: 978-3827410108, August 2002

[3-25] D. F. Caulfield, J. D. Passaretti, S. F. Sobczynski: Materials Interactions Relevant to the Pulp, Paper, and Wood Industries: 1990; 197 (MRS Proceedings) ISBN-13: 978-1558990869

[3-26] J.E. Holladay, J.J. Bozell, J.F. White, D. Johnson
Top Value-Added Chemicals from Biomass:Volume II — Results of Screening for Potential Candidates from Biorefinery Lignin
Prepared for the U.S. Department of Energy, October 2007
Contract DE-AC05-76RL01830

[3-27] M. Eggersdorfer: Perspektiven nachwachsender Rohstoffe in Energiewirtschaft und Chemie. Spektrum der Wissenschaft 1994; 6: 96-102

[3-28] F. S. Chakar, A. J. Ragauskas: Review of current and future softwood kraft lignin process chemistry. Industrial Crops and Products 2004; 20: 131-141

[3-29] R. R. Gustafson , C. A. Sleicher , W. T. McKean , B. A. Finlayson: Theoretical model of the kraft pulping process. Ind. Eng. Chem. Process Des. Dev. 1983; 22: 87-96

[3-30] H. Tanbdaa, J. Nakanoa, S. Hosoyab, H-M. Changc: Stability of α-Ether type Model Compounds During Chemical Pulping Processes. Journal of Wood Chemistry and Technology 1987; 7: 485-497

[3-31] A. Lindner, G. Wegener: Isolierung und chemische Eigenschaften von Ligninen aus Aufschlüssen nach dem Organocell-Verfahren und ihre potentielle nichtenergetische Verwertung. Das Papier 1988; 42: 1-8

[3-32] A. Lindner, G. Wegener: Characterization of Lignins from Organosolv Pulping According to the Organocell Process - Part 1. Journal of Wood Chemistry and Technology 1988; 8: 323-340

[3-33] A. Lindner, G. Wegener: Characterization of Lignins from Organosolv Pulping According to the Organocell Process - Part 2. Journal of Wood Chemistry and Technology 1989; 9: 443-465

[3-34] Lindner, G. Wegener: Characterization of Lignins from Organosolv Pulping According to the Organocell Process - Part 3. Journal of Wood Chemistry and Technology 1990; 10: 331-350

[3-35] P. N. Williamson: Repap's Alcell process: How it works and what it offers, Pulp and Paper Canada 1987; 12: 47-49

[3-36] E. K. Pye, J. H. Lora, The Alcell process: A proven alternative to kraft pulping, Tappi Journal 1991, 97: 113-118

[3-37] Axegård, P. Separation Processes in the Pulp Mill Biorefinery. In 1st Nordic Wood Biorefinery Conference, NWBC. 2008. Stockholm, March 11-13. p. 2-7.

[3-38] Öhman, F., Precipitation and Separation of Lignin from Kraft Black Liquor, 2006, Ph.D. Thesis, Chalmers University of Technology: Gothenburg, Sweden.

[3-39] Öhman, F., Theliander, H., Tomani, P., and Axegård, P., Method for Separating Lignin from Black Liquor. Patent WO2006031175, 2006.

[3-40] D. Feldman, D. Banu, M. Lacasse, J. Wang: Recycling lignin for engineering Applications. Materials research society symposiums proceedings 1992; 266: 177-192

4 Properties and Chemical Characterization

To produce superior grades of products from lignin, a detailed understanding of the properties and the chemical structure of lignin is necessary. Chapter 3.4 and 3.5 gave an overview concerning the anatomy of wood and a short introduction of the different kinds of wood, as well as the explanation of the Kraft pulping process, which is the most common process used in the paper mill industry. The organosolv process and a short description of processes in the bio refinery for the isolation of lignin from wood were shown in Chapter 3.5.3 and 3.5.4. The detection of the properties of lignin as a precursor for carbon fiber and the chemical characterization will now be shown in chapter 4.1 and 4.2.

4.1 Detection of the Properties of Lignin

To qualify lignin as a precursor for carbon fiber, a detailed characterization of the properties of lignin powder is necessary. Therefore thermo gravimetric analysis, differential scanning calorimetry is used to obtain these properties from Hardwood Lignin. All presented data of the analysis were performed on Hardwood Lignin, since this is the precursor used for carbon fiber production in this thesis. It is important to note, that all presented analyses in this chapter has been applied to Hardwood and Softwood Lignin. Chapter 4.1 gives an overview of important thermal properties of the Hardwood Lignin, which was used for the production of lignin based carbon fiber in laboratory scale. (compare Chapter 5)

The behavior of Hardwood Lignin during heat treatment is influenced by many factors. The understanding of these factors will help control the melt spinning process to suit Hardwood Lignin's behavior. Through thermo gravimetric analysis, the degradation of the Hardwood Lignin under heat treatment is detectable. Differential Scanning Calorimetry is used to analyze the glass transition temperature. The Combination of these results give insight to how the material can be used as a carbon fiber precursor.

4.1.1 Thermo Gravimetric Analysis

The thermo gravimetric analysis (TGA) detects the mass loss in a sample, correlating that to the temperature and/or the time. The loss of mass of a sample can be caused from different processes like chemical reactions, vaporization and degradation of the sample.

Theoretical Background

The change of the mass of the sample is measured with an electromagnetic compensating scale. Using the compensation signal in correlation to temperature and time, the mass loss is analyzed. The change in mass is normally published as absolute mass loss (mg) or relativity mass loss (%). An inert or oxidative stirring gas, which flows over the sample during measurement, is used to remove reaction or degradation products. Examples of inert gases are nitrogen and helium. Syntactic air and oxygen can be used as an oxidative gas. The general structure of a TGA is shown in Figure 34 [4-1].

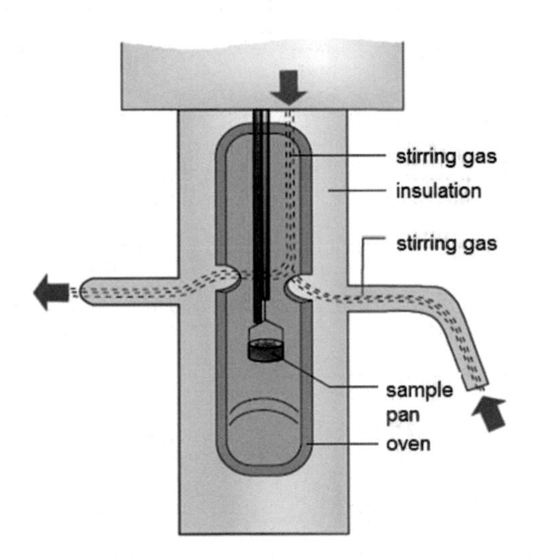

Figure 34: Principal structure of a TGA [4-1]

Experimental

The Thermo Gravimetric Analyzer Q500 from the company TA Instruments was used to make the TGA measurements. This analyzer works with the same principal as shown in Figure 34. The measurement range of this analyzer is from room temperature (23°C) up to 1000 °C, with a tolerance of ± 1 °C. To prevent any false measurements a sample pan made out of platinum was used.

For the measurements of Hardwood Lignin the samples were weighed in the platinum pan and then transferred in the TGA. All measurements were done in an inert nitrogen atmosphere. The TGA was heated at a rate of 5 K/min to reach a final temperature of 800 °C. The results were analyzed using a program named TA Universal Analysis.

Results and Discussion

The degradation of Hardwood Lignin consist of several chemical reactions that all take place during the analysis. The different thermal stability of the functional chemical groups, which can be found in Hardwood Lignin, causes the degradation of Hardwood Lignin at different temperatures (Figure 35 and 36). That is the reason for the degradation of Hardwood Lignin occurs over a wide temperature range [4-2].

The polymer structure of the Hardwood Lignin already begins to degrade at temperatures of 200 - 275 °C. From broken aryl ether bonds, highly reactive radicals are made which creates products with higher thermal stability. At around 330 °C the largest mass loss occurs due to the degradation. At this temperature acrylic and phenolic components were degraded. The degradation of Hardwood Lignin is also influenced by the moisture of the Hardwood Lignin sample, the used stirring gas, as well as heat and mass transport (Figure 35 and 36) [4-2].

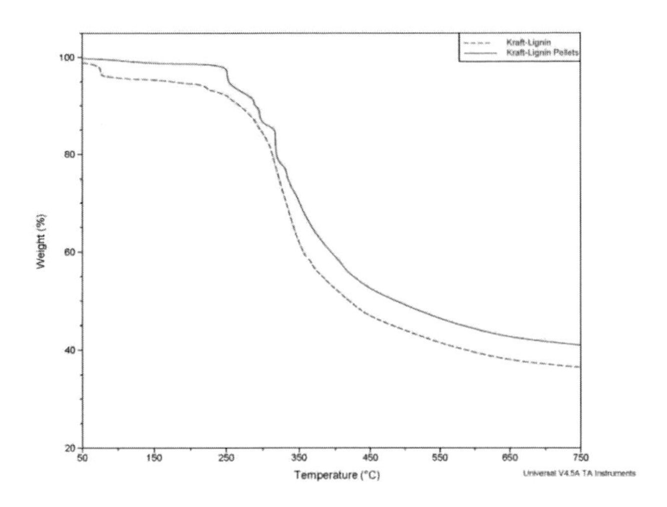

Figure 35: TGA measurement of Hardwood Lignin powder and pellets

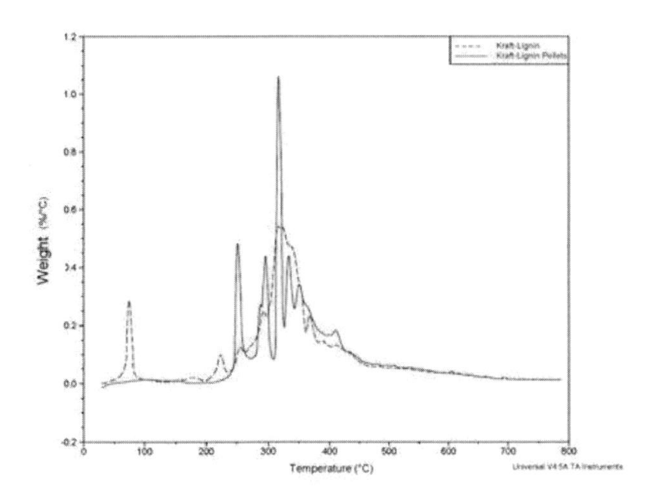

Figure 36: DTG measurement of Hardwood Lignin powder and pellets

The stabilized mass at the end of the degradation process, illustrates pure carbon and gives insight on the exceptionable conversion rate of Hardwood Lignin to carbon fiber.

4.1.2 Differential Scanning Calorimeter

Amorphous polymers like Hardwood Lignin have characteristic glass transition temperatures (T_g). At this temperature the solid state of the material changes to a viscoelastic state. The viscoelastic state can be analyzed with the help of a differential scanning calorimeter (DSC).

Theoretical Background

The DSC detects the heat quantity, which is necessary for physical effects or chemical reactions. The enthalpy is used to express the internal change of energy of a system at constant pressure. The enthalpy rises at endothermic processes like melting, vaporization, and glass transition. It drops down by exothermic processes like crystallization and degradation [4-3].

The glass transition temperature is influenced by the free volume between polymer backbones, the flexibility of the side chains, the length, and stability of the side chains. The degree of cross-linking, and change of chemical structures also influence the glass transition temperature. After reaching the glass transition temperature, the continuous heating leads to the melting of the material [4-4].

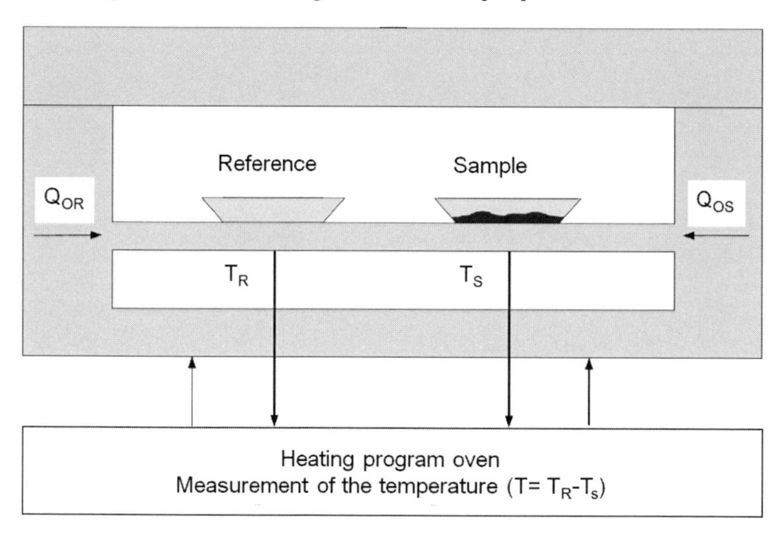

Figure 37: Schematically structure of the used DSC [4-3]

The DSC can operate in two different modes: dynamic heat flow differential calorimetry (Figure 37) and dynamic capacity differential calorimetry. Both modes measure calorically effects in comparison to a standard, which leads to compatible results [4-5].

Experimental

All measurements were done with the differential calorimeter Q2000 of the company TA Instruments. The calorimeter can operate in a temperature range of -90 °C to 400 °C. The DSC Q2000 works with the dynamic heat flow differential mode. The sample and the standard are in the same oven for measurement (Figure 37). All measurements were made with 7 to 8 mg of Hardwood Lignin. The heating rate was 10 K/min, and the sample was heated to 200 °C. This temperature was chosen from the results of the TGA measurements, since this is the point where the degradation of lignin starts. All measurements were made in inert nitrogen atmosphere.

Results and Discussion

The results of the DSC measurements are shown in Figure 38.

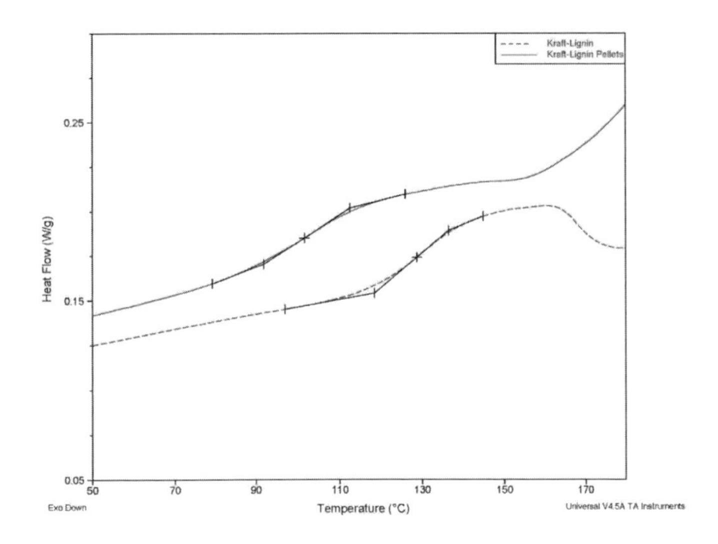

Figure 38: DSC measurements of hardwood Lignin powder and pellets

The TGA diagram shows the heat flow in connection to the temperature.

Table 7 summarizes the TGA data:

Table 1: DSC measurements of hardwood Lignin powder and pellets

	Onset temperature [°C]	**TG [°C]**	Endset temperature [°C]
Hardwood Lignin	119	**129**	134
Lignin Pellets	92	**102**	113

For the analysis of the results every experiment was done with two runs. The purpose of two runs has several advantages: During the first run water and moisture was eliminated and the contact between sample and pan was optimized [4-6].

The stable backbone of the polymer and crosslinking of the side chain leads to higher T_g. The variety of the molecular structure of the measured Hardwood Lignin (compare Chapter 4.3) finally leads to a glass transition temperature, which describes more of a range than a fixed temperature (Figure 38).

4.2 Chemical Characterization of Lignin

To qualify lignin as a precursor for carbon fiber, a detailed chemical characterization of this new material is necessary. Therefore elementary analysis (detection of the C9-units), mass spectroscopy (analysis of molecular fragmentation), nuclear magnetic resonance spectroscopy (analysis moleculare structure) and Fourier transform infrared spectroscopy (analysis chemical structure) are used.

Chapter 4.3 gives an overview for the most important chemical properties and structures of the hardwood lignin which is used for the production of lignin based carbon fiber in laboratory scale shown in Chapter 5.

The chemical structure will be defined as follows:

1. Elementary analysis for detection of the C9-units
2. mass spectroscopy for analysis of defragmentation and side-chains
3. nuclear magnetic resonance spectroscopy and Fourier transform infrared spectroscopy for complete analysis of chemical bonding in the lignin macromolecule and defining of the chemical structure

4.2.1 Elementary Analysis

The elementary analysis is a technique for the determination of the weight percent of chemical elements in organic components. From this data and the molar mass determination it is possible to investigate the empirical formula. This way the stoichiometry of an organic compound like lignin becomes available. Through further physical analysis it is also possible to find the structure formula [4-7].

Theoretical Background

The most common form of elemental analysis, CHN analysis, is accomplished by combustion analysis. In this technique, the sample is exactly weighted and then catalytically burned at high temperatures (1800°C using exothermic reactions) with a pure oxygen environment. Then the gases from the combustion (oxidation products) are transferred to a catalytic bed, made from copper or wolfram, with the aid of an inert gas (in most cases helium). Nitrogen oxides are reduced over this catalytic bed into pure nitrogen at temperatures of 600 - 900°C. After that, the gases (CO_2, H_2O, SO_2 and N_2) are separated into separation columns (adsorption and desorption separation column), and quantitative detected by the thermal conductivity detectors (TCD). Using this data, the composition of the unknown sample is calculated [4-8].

Experimental

The investigations of the contained elements in the used Hardwood Lignin are done with the Vario ELcube CHNS from Elementar. The sulfur-carbon-analyzer CS230 and the CHN-analyzer from LECO were used. The water content was measured with the water detector Ma 45 from LECO. Each test was performed until the average for the lignin was stabilized and the results are shown below.

Results and Discussion

For the Hardwood Lignin powder a detailed elementary analysis was made (Table7).

Table 7: Results of the elemental analysis of the used Hardwood Lignin powder

Analyzing Tool	Element	Content	RSD
CHN628	C	61,73%	0.423
	H	6,1%	0.668
	N	0,12%	14.19
Ma 45	H_2O	4,5%	0.886
CS 230	S	3,2%	0.927

With the assumption that the lignin monomer is assembled only from the elements carbon, hydrogen and oxygen, the oxygen ratio was estimated as the difference to 100%. For the creation of the C_9-unit the sulfur contamination was not used, since sulfur is a chemical used for the Kraft pulping and is not part of the lignin structure. Using the carbon, hydrogen, and oxygen data, the C_9-unit was calculated with: $C_9H_{10,8}O_{3,56}$. The formula C_9 contains the complete information about the lignin monomer structure (Figure 39). For examples of the C_9-units of lignin also compare chapter 3.4 (Table 4: Monomers of Lignin).

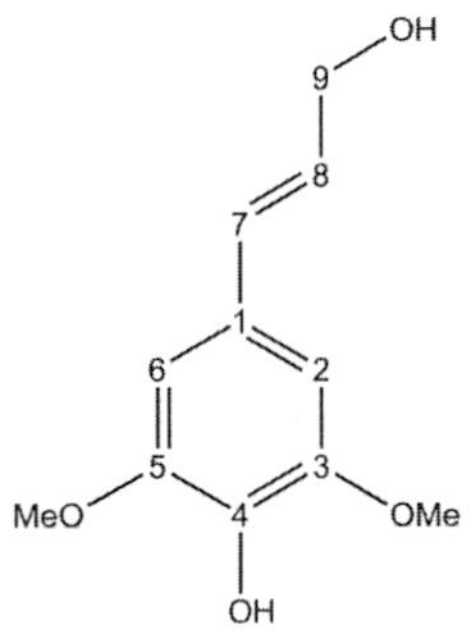

Figure 39: Sinapyl alcohol as an example for C_9-units

The elementary analysis shows, that Hardwood Lignin has a carbon content of over 60% (see Table 7). That means a theoretical conversion rate for making carbon fiber from lignin of over 60% would be possible. The experimental data shows that the conversion rates are much lower at 35% (compare Chapter 5).

4.2.2 Mass Spectroscopy

Mass spectrometer allows quantitation of atoms or molecules and provides structural information by the identification of distinctive fragmentation patterns. This is accomplished by using the difference in the mass-to-charge ratio (m/z) of ionized atoms or molecules in ultra-high vacuum environment [4-9].

Theoretical Background

The general operation of a mass spectrometer is shown in Figure 40. There are three major steps that are important for this measurement [4-10]:

1. create gas-phase ions
2. separate the ions in space or time based on their mass-to-charge ratio
3. measure the quantity of ions of each mass-to-charge ratio

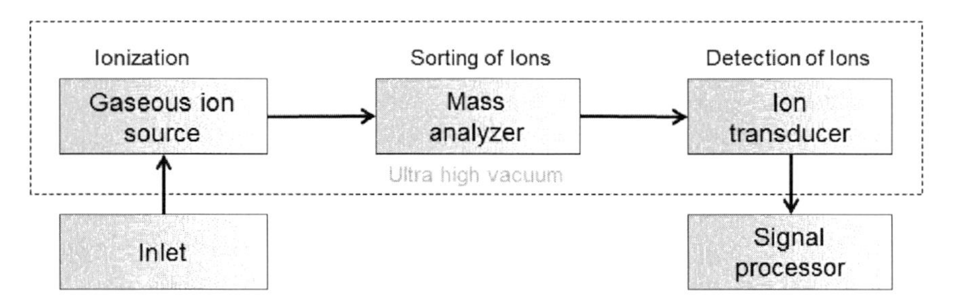

Figure 40: Schematic structure of a mass spectroscope

Experimental

The investigation of the Hardwood Lignin powder was made with a high-resolution mass spectrometer (sector field instrument) by the company Finnigan. The following conditions were used:

- temperature-controlled direct steam generation of the sample
- electron impact and chemical-ionization-energy of 50eV
- detection of molecular mass between 50 and 500u

Results and Discussion

The profile of volatilization of the Hardwood Lignin powder shows the maximum fragmentation (relative abundance = 100) at around 1 minute. This area has the highest volatilizing intensity. The next largest abundance occurs at 1.5 minutes approximately (relative abundance = 45). The received spectra (Figure 42, Figure 43 and Figure 44) will be discussed in an overall perspective to analyze all possible fragmentation products.

Figure 41 shows the profile of volatilization of the used Hardwood Lignin powder.

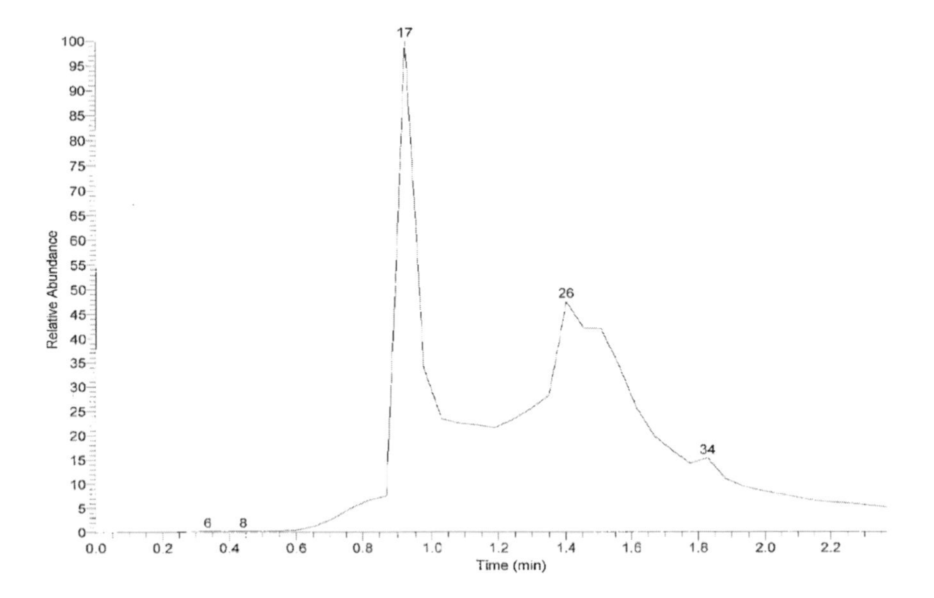

Figure 41: Profile of volatilization of Hardwood Lignin

Figure 42, 43, and 44 show the received spectra of the mass spectroscopy for the Hardwood Lignin powder.

The measurement range from 50 to 500u was chosen, to detect all possible fragmentation products. Figure 45 and 46 illustrate the monomers and dimeres, which makes up the defragmentation products.

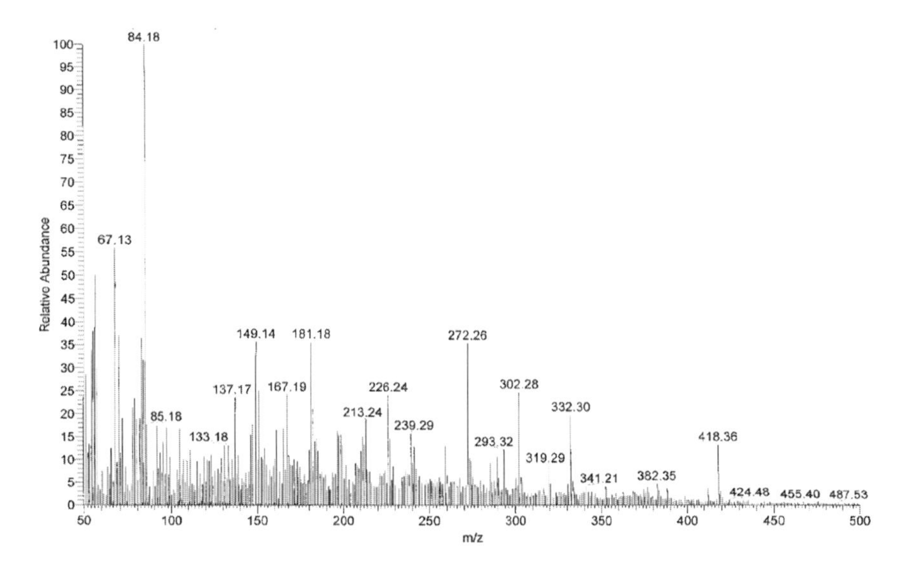

Figure 42: Mass spectroscopy spectra of Hardwood Lignin (overview spectrum)

It is nessessary to control the temperature to ensure that the transformation of the solid powder into the gaseous phase occurs. The area of the capital evaporation can be found after the one minute mark, at a temperature of 120°C and around 10^{-8}mbar. In this area the monomeres (Figure 41) and their fragmentation products were found. The molecular weight of the monomeres (H, G, S and S') are in the range of 124u and 182u. The fragmentation of the monomeres leads to the molecular weight under the weight of the monomeres. (Figure 45 and 46). Below are the fragmentation steps and their respective masses:

- Δm=14u (elemination of CH_2)
- Δm=15u (elemination of CH_3)
- Δm=16u (elemination of oxigen)
- Δm=31u (elemination of O-CH_3)

The peaks at the higher molecular weights show dimeres of the Hardwood Lignin and their fragmentation products. Typical Δm for couplings like C-O-C bondings or C-C bondings are an indicator of how the monomeres are coupled to dimeres and threemeres. A detailed analysis of the spectra led to the Hardwood Lignin monomeres and dimeres shown in Figure 43 and 44.

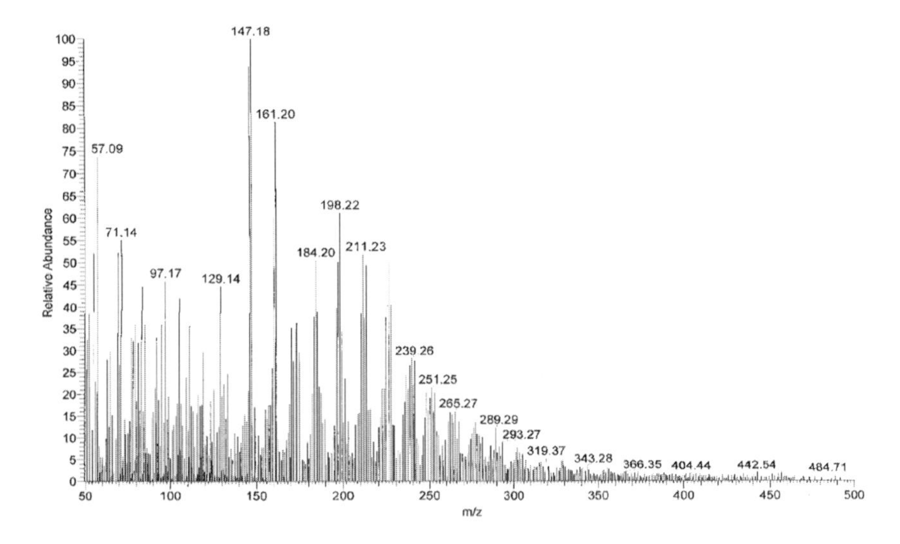

Figure 43: Mass spectroscopy spectra of Hardwood Lignin (Lignin monomers)

The measurement of the molecular weights of the dimeres and their fragmentation products can be found after 90 seconds of measurement time, at a temperature of 180°C and around 10^{-8}mbar. The spectrum of this evaporation area (Figure 44) illustrates that. The chemical structure of the dimeres can be found in Figure 46.

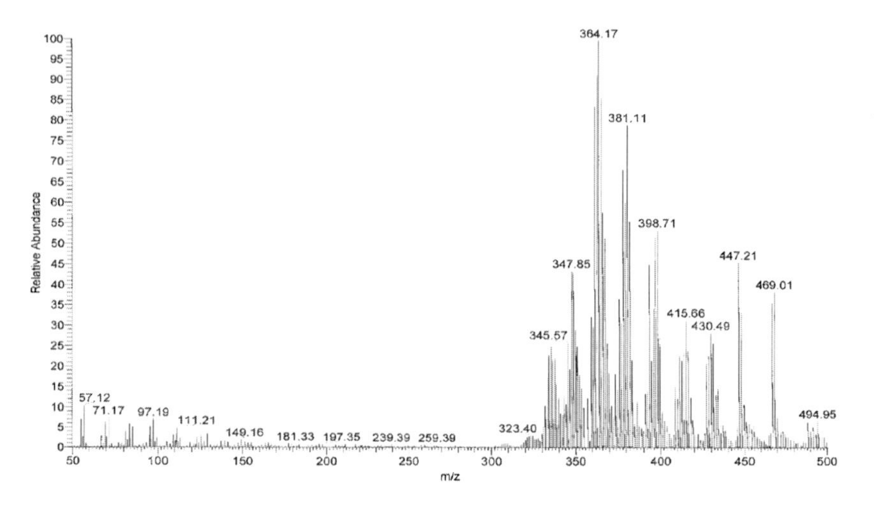

Figure 44: Mass spectroscopy spectra of Hardwood Lignin (Lignin Side-Chains)

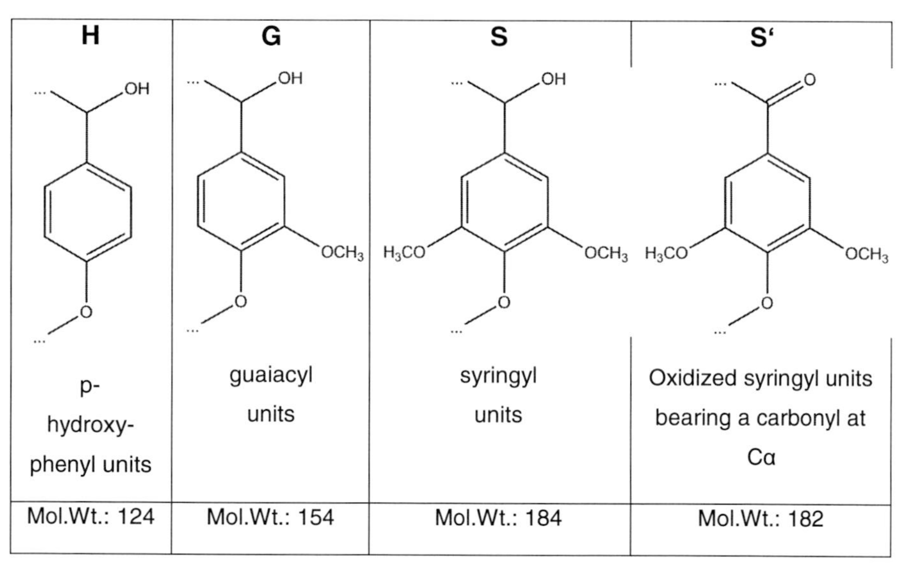

H	G	S	S'
p-hydroxy-phenyl units	guaiacyl units	syringyl units	Oxidized syringyl units bearing a carbonyl at Cα
Mol.Wt.: 124	Mol.Wt.: 154	Mol.Wt.: 184	Mol.Wt.: 182

Figure 45: Hardwood Lignin monomers and their molecular weight

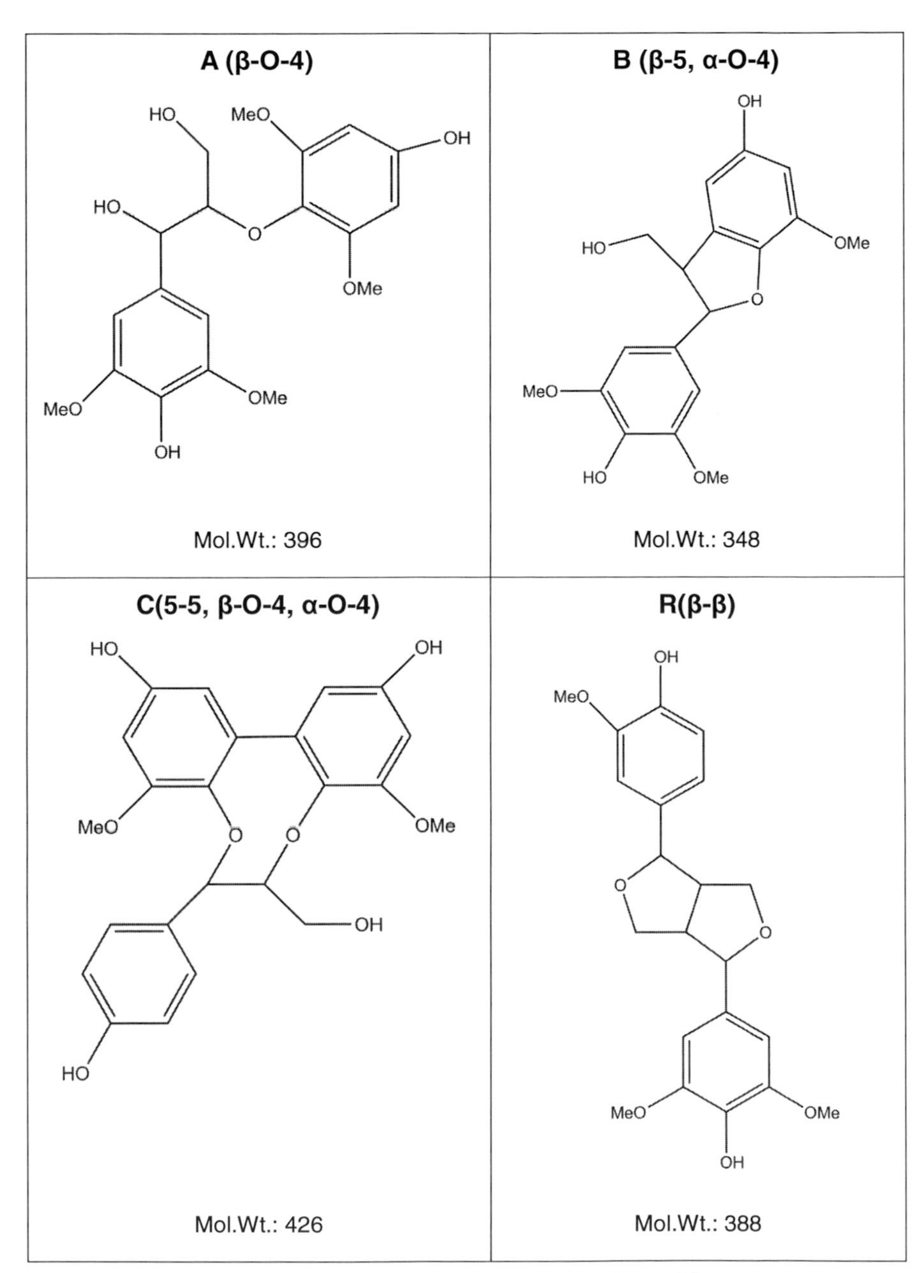

Figure 46: Hardwood Lignin side-chains and their molecular weight

Detailed structural analysis using mass spectroscopy

Several structural parts of the Hardwood Lignin, which were used for the carbon fiber production in laboratory scale, were established with mass spectroscopy.

Since sinapyl alcohol is the main component of Hardwood Lignin, the monomers syringyl units (named S in Figure 47) and the oxidized syringyl units bearing a carbonyl at Cα (named S' in Figure 47) are the dominating structures. They can be found in all spectra as a monomer or as part of dimeres (Figure 47) with the molecular weight of 184u for the syringyl units and 182u for the oxidized syringyl units.

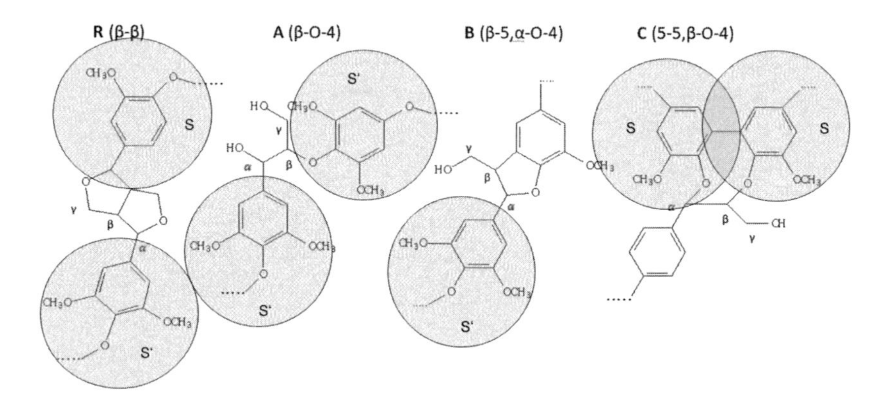

Figure 47: Presence of syringyl units and *oxidized syringyl units bearing a carbonyl at Cα in Hardwood Lignin side chains*

Table 8 gives an overview of the spectra syringyl units (named S in Figure 47), and the oxidized syringyl units bearing a carbonyl at Cα (named S' in Figure 47) can be detected.

Also the complete side-chains and the derivatives of A, B, and C (Figure 48) are detectable (Table8).

The nomenclature for the derivatives of A, B, and C can be found in Figure 46. The phenyl derivatives are shown in Figure 49.

Table 8: Detection of syringyl units and *oxidized syringyl units bearing a carbonyl at Cα with mass spectroscopy*

Unit	Peak	Spectrum
C_3	57,09	FIgure 43
P_1	67,13	Figure 42
P_2	84,18	Figure 42
P_3	147,18	Figure 43
B_2	149,14	Figure 42
P_4	161,20	Figure 43
C_4	181,18	Figure 42
B	347,85	Figure 44
C_2	272,26	Figure 42
S' as part of B	347,85	Figure 44
Condensation of 2 S'	364,17	Figure 44
C (derivate without OH / OME)	381,11	Figure 44
A	398,71	Figure 44
S' as part of A	398,71	Figure 44

In summary, the mass spectroscopy is a very useful tool for analyzing Hardwood Lignin. This tool makes it possible to characterize the major chemical structural elements. The most important structural elements for making a carbon fiber are the phenolic rings, which will create the graphite structure in the carbon fiber. These phenolic rings were successfully detected.

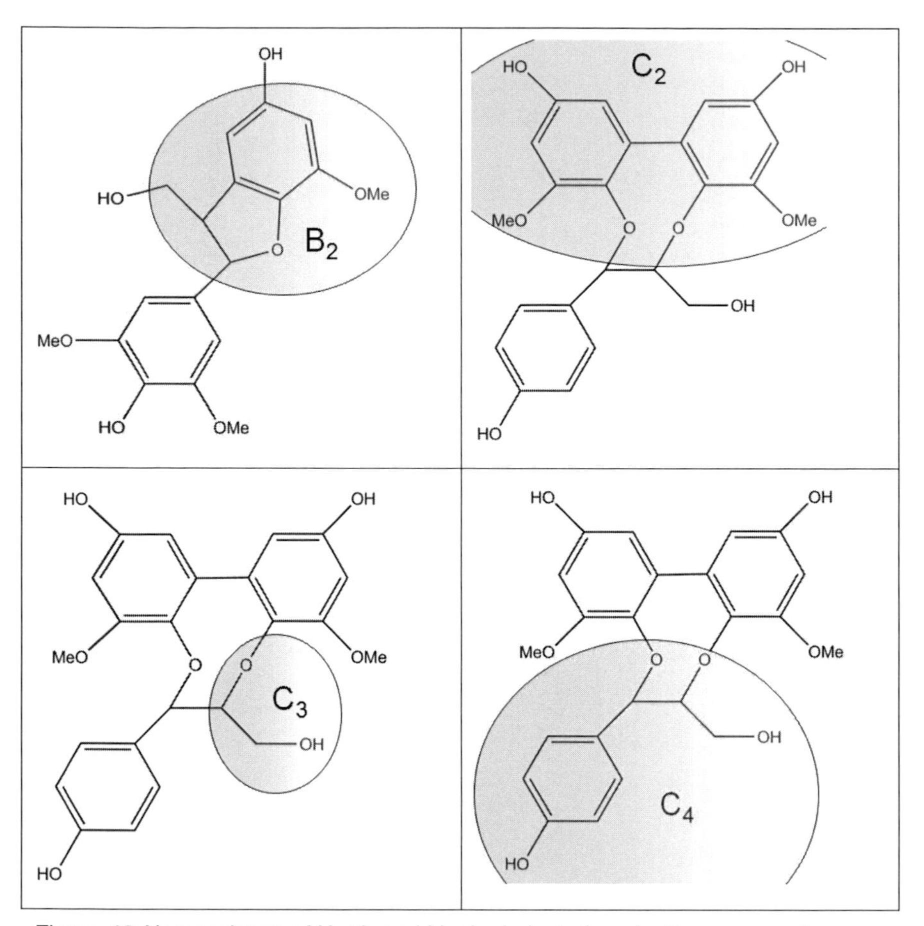

Figure 48: Nomenclature of Hardwood Lignin derivate found with mass spectroscopy

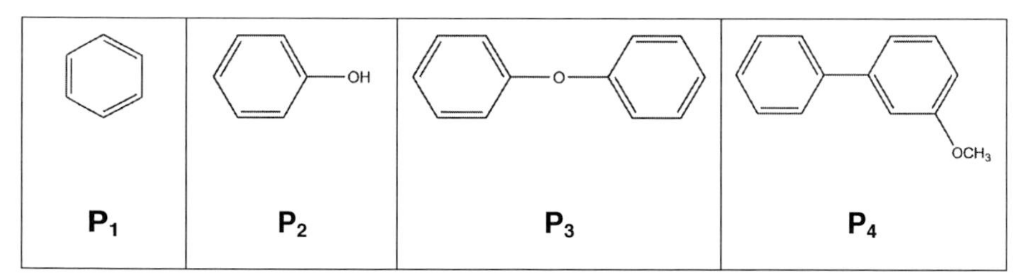

Figure 49: Nomenclature of Hardwood Lignin phenyl-derivate found with mass spectroscopy

4.2.3 Nuclear Magnetic Resonance Spectroscopy

Nuclear Magnetic Resonance (NMR) Spectroscopy has the ability to differentiate subtle differences in molecular configurations, which makes it an ideal technique for identifying monomers and side chains of Hardwood Lignin [4-11].

Theoretical Background

The nucleus of every atom has a specific rotation about their axis, known as nuclear spin. The nucleus of the isotope ^{1}H for example is positively charged and the rotation creates a magnetic moment. By using an external magnetic field an orientation of the nucleus is possible. In a homogeneous magnetic field the nucleus can be found in defined levels. For transferring the nucleus from a lower to a higher level of energy the nucleus needs to adsorb or desorb the energy difference, which means it needs to adsorb or desorb radiation of the exact amount of energy. This effect is called spin reversal. The corresponding energy is adsorbed and measured with the help of the Nuclear Magnetic Resonance Spectrometer. NMR focuses on monitoring the behavior of nuclear magnetic activity in a magnetic field. NMR measures the amount of energy which the nuclei absorb in a homogeneous magnetic field. [4-11, 4-12, 4-13, 4-14, 4-15]

Experimental

The one-dimensional (1D) and two-dimensional (2D) NMR spectra (^{1}H and ^{13}C) were measured with a Bruker AVANCE 600 NMR spectrometer. The operating frequency was 600.13 MHz for ^{1}H spectrum and 150.9 MHz for ^{13}C spectrum. The temperature was constant at 295 K (±0.1 K), as the solvent DMSO-d_6 was used. The solvent was also used as an internal standard for the referencing of the chemical shift of ^{1}H (2.49 ppm) and ^{13}C (39.5 ppm). For classification of the chemical shift measurements of two-dimensional NMR (COSY and HSQC) were made.

COSY (COrrelated SpectroscopY) experiments

The step size in the COSY experiment was 6 kHz in both dimensions. They were measured in increments of 256 and 2k data points. The number of scans was 32 and the pulse width for ^{1}H was 13.0 µs.

HSQC (Heteronuclear Single Quantum Coherence) experiments

The HSQC experiments were carried out with a Bruker AVANCE 600 NMR spectrometer operating at 600.13 MHz and 298 K. The spectrometer was fitted with a 5 mm TBI-^{1}H-^{13}C/^{15}N/^{2}H probe head with z-gradients. The ^{1}H and ^{13}C chemical shifts were determined relative to the internal standard DMSO-d_6 and were given in parts per million downfield to TMS (Tetramethylsilane).

The HSQC experiments were separately measured for the aliphatic and for the aromatic regions with a spectral width in F2 of 1.3 kHz and 2.7 kHz, respectively. For all experiments, the spectral width in F1 was 8 kHz. The HSQC experiments were detected using 2 K data points in F2 with 256 scans and 128 scans in F1. The relaxation delay was 1.5 seconds and pulse width for ^{1}H was 9.3 µs and for ^{13}C was 13.0 µs.

^{31}P NMR spectra

The ^{31}P NMR spectra were obtained by using inverse gated decoupling on a Bruker AVANCE 400 NMR spectrometer, operating at 161.9 MHz and using a 5 mm BBOF probe head with z-gradients. The external standard used was 85% H_3PO_4. The lignin samples were dissolved in 0.75 ml of a mixture of deuterated pyridine and chloroform in ratio of 3:2 as solvent, with a 0.47 mol/l cyclohexanol solution as an internal standard (145 ppm). The samples were simultaneously derived in the NMR tube with 0.1 ml 2-chloro-4,4,5,5-tertamethyl-1,2,3-dioxaphospholane (TMDP). The inverse gated decoupling sequence was used with a 25 second relaxation delay and 128 scans were collected.

Solid State NMR

The ^{13}C CP/MAS spectra were obtained on a Bruker Avance 300 NMR spectrometer, operating at 75.47 MHz. The spectrometer was equipped with 4 mm CP/MAS probe-head BBO/ML4. The samples were measured in 4 mm ZrO_2 MAS rotors. The proton 90° pulse was set to 2.55 µs and the decoupling frequency during acquisition was 54 kHz. Following conditions were applied: 5 kHz spinning speed, recycle delay 3 s, contact time 1.74 ms. Chemical shifts are quoted in ppm from TMS. The samples were pulverized and mixed with an equal weight of hydrated magnesium silicate, to reduce the macro-current generation under radio frequency excitation.

Allocation Strategy

The 2D NMR spectroscopy was used to identify the primary structures of the investigated Hardwood Lignin derivatives. The assignments of the ^{1}H chemical shift and ^{13}C chemical shift are based on the HSQC (Heteronuclear Single-Quantum Correlation) and HMBC (Heteronuclear Multiple Bond Correlation) experiments.

The spectra can divided in three parts – the aromatic, side-chain and aliphatic (Figure 50 and Figure 51) [4-11, 4-12, 4-13].

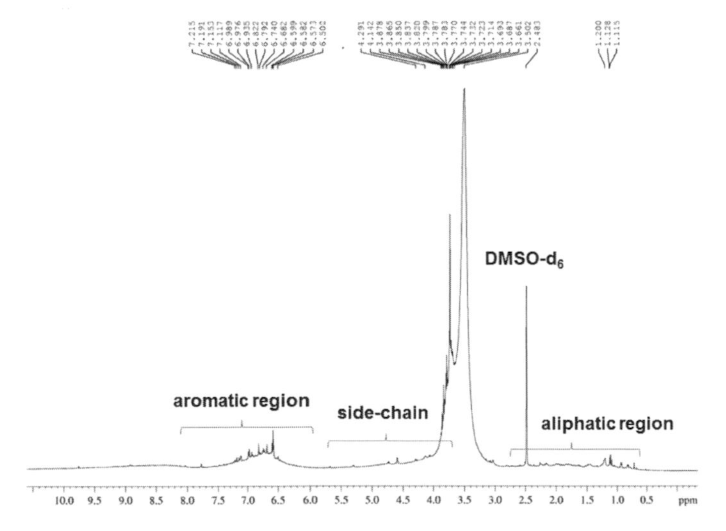

Figure 50: ^{1}H NMR Spectrum of Hardwood Lignin, measured in DMSO-d_6

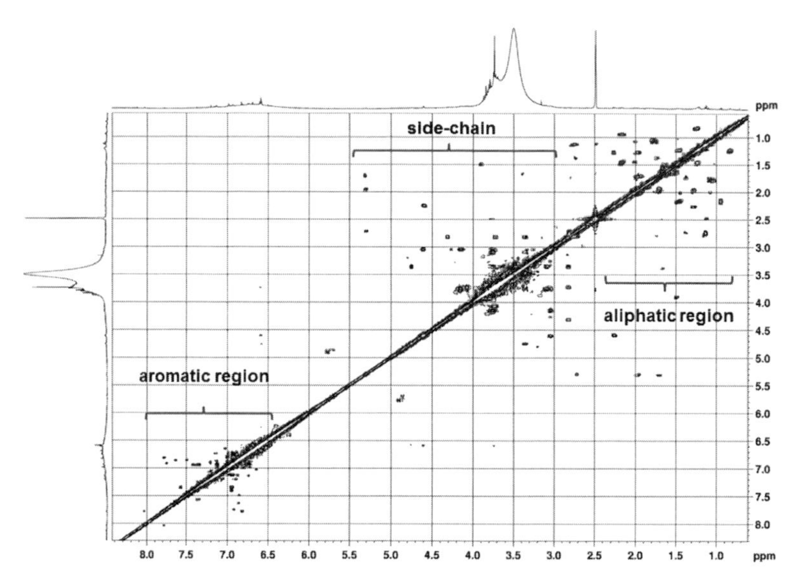

Figure 51: $^{1}H,^{1}H$-COSY of Hardwood Lignin, measured in DMSO-d_6

The signals in aliphatic regions are not sensitive for structural modifications. For the structure elucidation the aromatic and side-chain regions are of particular interest to identify the Hardwood Lignin monomers and inter-unit linkages [4-11, 4-12, 4-13].

Important information concerning the structure and composition of Hardwood Lignin were detected with the help of the ^{13}C chemical shift in the aromatic region (100-150 ppm) and the side-chain region (50-90 ppm).

Results and Discussion

The 2D NMR spectroscopy was used to identify the primary structures of the investigated Hardwood Lignin derivatives. The assignments of ^{1}H and ^{13}C chemical shifts are based on the HSQC (Heteronuclear Single-Quantum Correlation) and HMBC (Heteronuclear Multiple Bond Correlation). The spectra can be divided in to three major parts – the aromatic, side-chain and aliphatic part. The signals in aliphatic region are not sensitive for structural modifications. For the structure elucidation, the aromatic, and side-chain regions are of particular interest to identify the Hardwood Lignin monomers and inter-unit linkages (side-chains). The HSQC and HMBC spectrums for the Hardwood Lignin were presented in Figure 54 and 55. For every sample, the aromatic, and the side-chain region were separately scanned. The cross-peaks were assigned by comparing the literature [4-14, 4-15, 4-16, 4-17, 4-18] and are listed in Table 9 and Table 10. For the assignments compare Figure 52 and Figure 53.

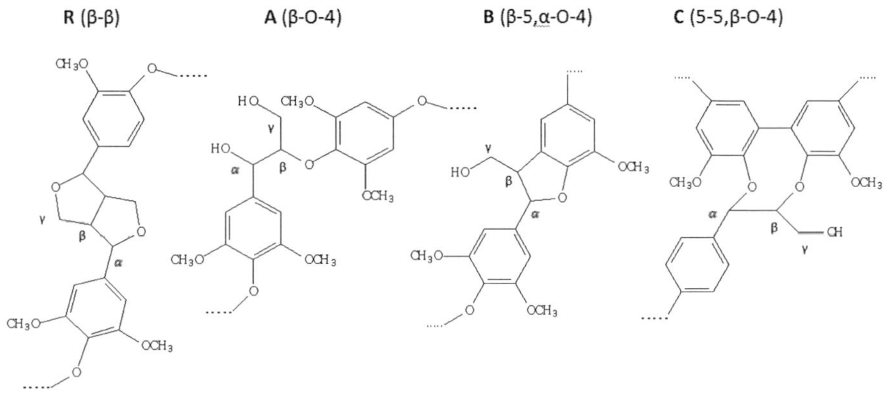

Figure 52: Hardwood Lignin side chains

The investigated samples were produced from Hardwood Lignin powder, which typically contain guaiacyl **G** and syringyl **S** units (Figure 55).

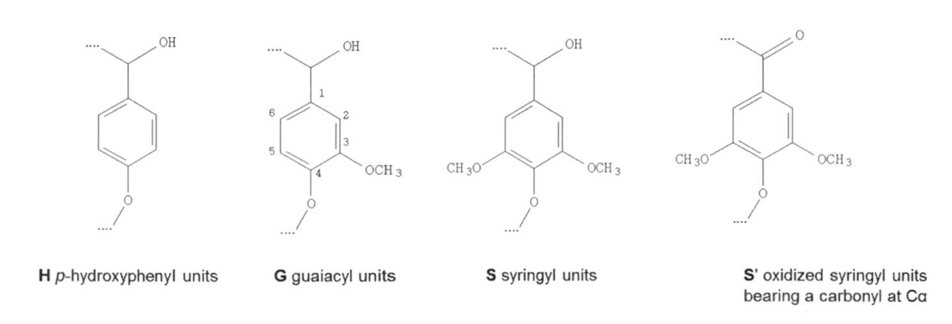

Figure 53: Lignin Monomers

The side-chain region is ranging from 50 - 90 ppm for ^{13}C and from 2.0 – 5.0 ppm for ^{1}H chemical shifts (Figure 54 and Table 9).

The methoxy group, the main signal, can be located at δ_C/δ_H 56.3/3.74 ppm. In the range of δ_C/δ_H 71.0-72.0 and 4.75-4.85 ppm the signals for substructure $\mathbf{A}_\alpha$ (C_α-H_α) can be found. $\mathbf{A}_\beta$ (C_β-H_β) is linked to **G** and **S** and is located at 81.5/4.75 ppm. The signal for $\mathbf{A}_\gamma$ is located at 87.7/4.60 ppm (Figure 55).

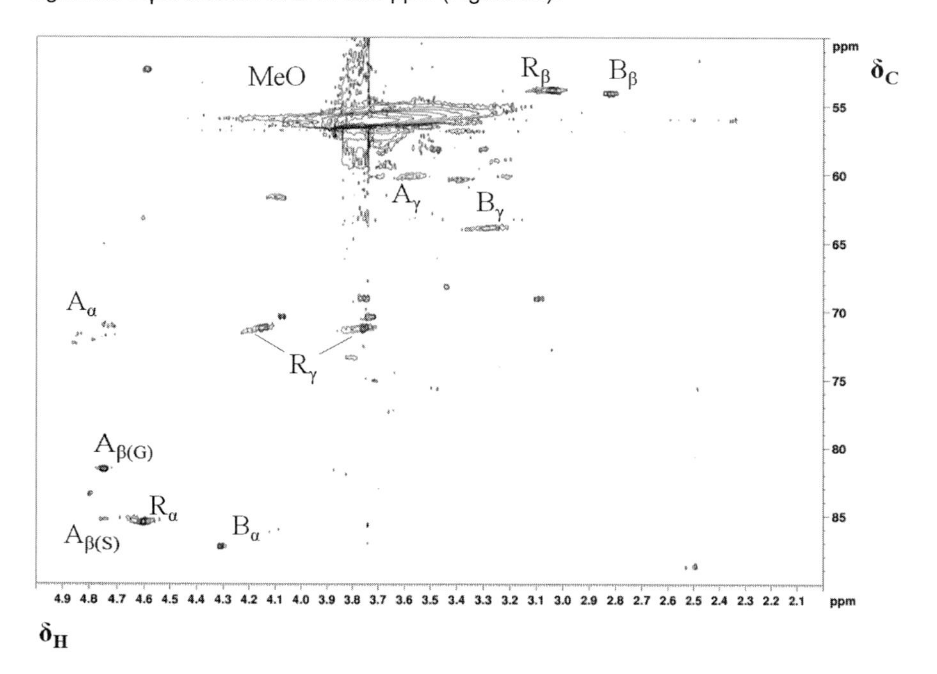

Figure 54: HSQC spectra for the side-chain region of Hardwood Lignin

For C_γ-H_γ, the cross peaks were observed at δ_C/δ_H 60.2/3.57 ppm. Furthermore, strong signals were observed for resinol substructure **R** at δ_C/δ_H 54.0/3.05 ppm for C_β-H_β, 71.4/4.75 and 3.37 ppm for C_γ-H_γ and 87.7/4.60 ppm for C_α-H_α. (Figure 46) The signals for phenylcoumaran **B,** an already known inter-unit linkage in lignin, were detected at δ_C/δ_H 54.3/2.83 ppm for C_β-H_β, 64.1/3.27 ppm for C_γ-H_γ and 87.4/4.30 ppm for C_α-H_α. (Figure 54)

Table 9: Assignment of ^{13}C-^{1}H correlation signals of side-chain region in the HSQC spectra of Hardwood Lignin powder

Label	^{13}C in ppm	^{1}H in ppm	Assignment
R_β	54.0	3.05	C_β in **R**
B_β	54.3	2.83	C_β in **B**
O-CH_3	56.3	3.74	**O-CH_3**
A_γ	60.2	3.57	C_γ in **A**
A_γ	61.0	4.12	C_γ in **A**
A'_γ	-	-	C_γ in **A'**
B_γ	64.1	3.27	C_γ in **B**
R_γ	70.4	4.08 / 3.73	C_γ in **R**
R_γ	71.4	4.16 / 3.75	C_γ in **R**
A_α	71.0	4.75-4.85	C_α in **A** linked to **G**
A_α	72.0	4.75-4.85	C_α in **A** linked to **S**
A_β	81.5	4.75	C_β in **A** linked to **G**
R_α	85.7	4.60	C_α in **R**
A_β	85.7	4.60	C_β in **A** linked to **S**
B_α	87.4	4.30	C_α in **B**

The aromatic region is ranging from 100 - 150 ppm for ^{13}C and from 5.5 – 7.5 ppm for ^{1}H chemical shifts (Figure 55 and Table 10).

The syringyl **S** and guaiacyl **G** were detected in the aromatic region, ranging from 100 – 140 ppm for ^{13}C and 5.0 -7.5 ppm for ^{1}H. The **S** units showed different correlations for $C_{2/6}$-$H_{2/6}$ δ_C/δ_H 103.4-105.3/6.58-6.87 ppm depending on the different inter-unit linkages at C_4 and C_1. The C_α-oxidized **S** units (C2/6-H2/6) were found at δ_C/δ_H 106.6-108.3/7.12-7.22 ppm. The **G** units revealed dominant signals (C_2-H_2) at δ_C/δ_H 109.6/7.12, and the broad signal (C_5-H_5) at δ_C/δ_H 115.2-115.5/6.6-6.8 ppm. A weaker C_6-H_6 signal was detected at δ_C/δ_H 119.6/6.93 ppm. Signals for **H** were detected at δ_C/δ_H 126.2-126.5/6.82-6.93 ppm. (chemical structure comparison, Figure 53)

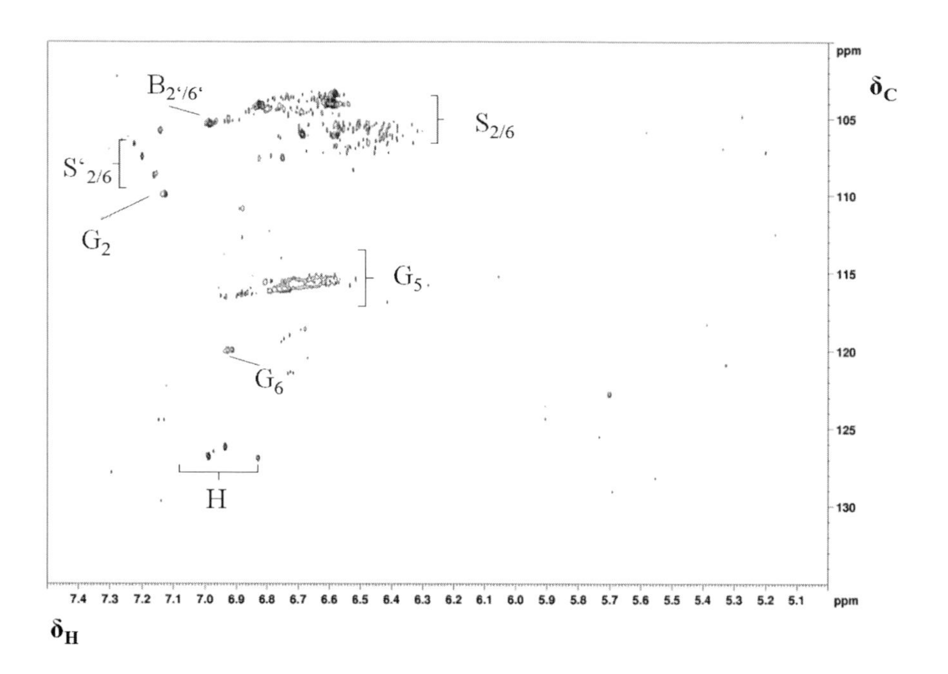

Figure 55: HSQC spectra for the aromatic region of Hardwood Lignin

Table 10: Assignment of ^{13}C-^{1}H correlation signals of aromatic region in the HSQC spectra of Hardwood Lignin

Label	**^{13}C in ppm**	**^{1}H in ppm**	**Assignment**
$S_{2/6}$	103.4	6.58	$C_{2/6}$-$H_{2/6}$ in **S** linked to **A**
$S_{2/6}$	103.7	6.59	$C_{2/6}$-$H_{2/6}$ in **S** linked to **R**
$S_{2/6}$	103.7	6.61	$C_{2/6}$-$H_{2/6}$ in **S** linked to **B**
$S_{2/6}$	103.7	6.83	$C_{2/6}$-$H_{2/6}$ in **S**
$S_{2/6}$	105.3	6.98	$C_{2/6}$-$H_{2/6}$ in **B**
$S_{2/6}$	105.3	7.13	$C_{2/6}$-$H_{2/6}$ in **S**

$S_{2/6}$	105.3	6.69	$C_{2/6}$-$H_{2/6}$ in **B**
$S'_{2/6}$	106.6	7.22	$C_{2/6}$-$H_{2/6}$ in oxidized **S**
$S'_{2/6}$	107.5	7.19	$C_{2/6}$-$H_{2/6}$ in oxidized **S**
$S'_{2/6}$	108.3	7.16	$C_{2/6}$-$H_{2/6}$ in oxidized **S**
G_2	109.6	7.12	C_2-H_2 in **G**
G_5	115.6	6.6-6.6	C_5-H_5 in **G**
G_5	115.2	6.6-6.8	C_5-H_5 in **G**
G_6	119.6	6.93	C_6-H_6 in **G**
$H_{2/6}$	126.2	6.93	$C_{2/6}$-$H_{2/6}$ in **H**
$H_{2/6}$	126.4	6.82	$C_{2/6}$-$H_{2/6}$ in **H**
S_1	133.6-135.4	-	C_1 in S_1
S'_1	140.0-141.8	-	C_1 in S'_1
S_4, $S_{3/5}$ G_3, G_4	147.1-148.2	-	C_4, $C_{3/5}$ in etherified S C_3, C_4 in etherified G
	128.1/129.5	-	
C=O	n.o.	-	C=O in acetates
C=O	190.8 – 196.3	-	α-CO / γ-CHO

The efficiency of NMR as an analytical technique for the characterization of Hardwood Lignin can be confirmed by the data collected from the HBMC tests (Figure 56). This spectra allowed for the assignment of quaternary carbons C_1, $C_{3/5}$, and C_4 in **S** units, the carboxyl groups in oxidized syringyl groups in **S'**, and the C_1, C_3, and C_4 in **G** units. Strong long range correlations for the aromatic protons in **S** and **G** units were observed between $H_{2/6}$, and H_5, as well as the inter-unit linkages C_α in **B,** and **R,** C_β in **A** and **B** are dominant. (chemical structure, Figure 52 and 53)

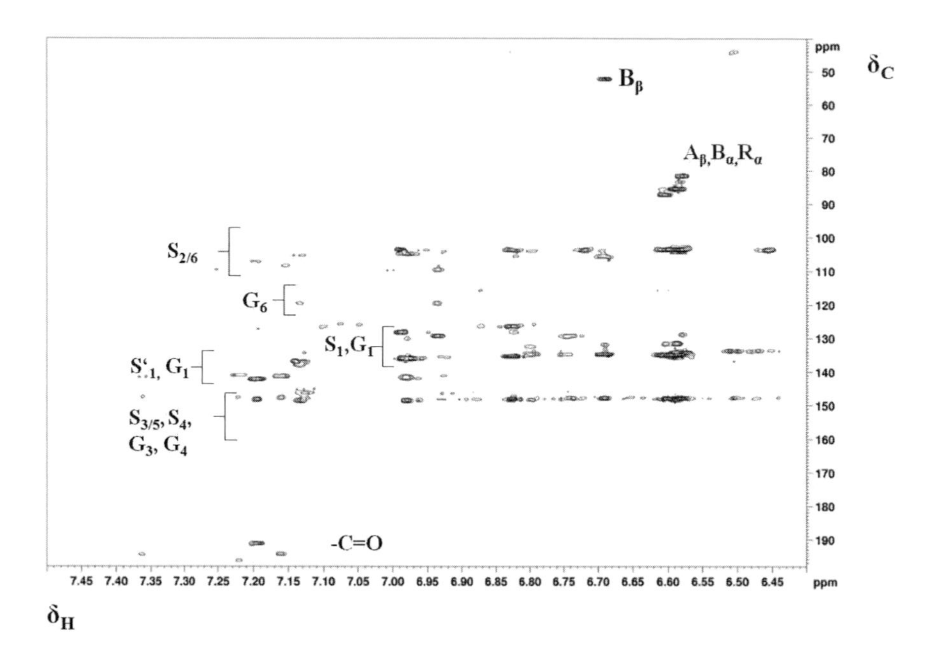

Figure 56: HMBC spectra for the aromatic region of Hardwood Lignin

The utilization of 2D NMR for the characterization of Hardwood Lignin is a useful tool for the characterization of the chemical structure of Hardwood Lignin. The results show the complexity of lignin as a natural product. The enormous variety of chemical bonds in the lignin macro molecule, make it very difficult to predict the chemical reactions during the conversion process (shown in Chapter 5 - Development of a Lignin Based Carbon Fiber). Chapter 6 (Reaction Mechanism of Lignin during Conversion to Carbon Fiber) will explain the major chemical reactions during this conversion process.

4.2.4 Fourier Transform Infrared Spectroscopy

Infrared (IR) Spectroscopy is the analysis of infrared light interacting with a molecule. The infrared region of the electromagnetic spectrum is light with a longer wavelength and lower frequency than visible light. The IR Spectroscopy covers a range of techniques, mostly based on absorption spectroscopy. IR Spectroscopy measures the vibrations of atoms to determine the functional groups of Hardwood Lignin [4-19].

Theoretical Background

Infrared Spectroscopy is used to determine the structures of molecules and the molecules' characteristic absorption of infrared radiation. The infrared spectrum, along with ultraviolet and Raman spectrums, is one of the molecular vibrational spectrums. When exposed to infrared radiation, sample molecules selectively absorb radiation of specific wavelengths which causes the change of dipole moment of the sample molecules. Consequently, the vibrational energy levels of the sample molecules transfer from ground state to excited state. The frequency of the absorption peak is determined by the vibrational energy gap. The number of absorption peaks is related to the number of vibrational freedoms of the molecule. The intensity of the absorption peaks are related to the change of the dipole moment and the possibility of the transition of energy levels. Therefore, by analyzing the infrared spectrum, it is possible to obtain abundant structure information of Hardwood Lignin [4-19, 4-20].

Fourier transform infrared (FTIR) spectroscopy is a measurement technique that allows to record infrared spectra. A data-processing technique called Fourier transform turns this raw data into the desired result (the sample's spectrum): Light output as a function of infrared wavelength (or equivalently, wavenumber) [4-19, 4-20].

Experimental

The Infrared spectrums were recorded on a Perkin Elmer FT/IR 2000. The samples were measured using an ATR unit (attenuated total reflection). The spectral resolution was 4 cm^{-1}. The amount of samples used during IR measurements was kept constant. To ensure a full analysis of Hardwood Lignin scans for IR bands from 4400 to 400 cm^{-1} were performed, although the IR bands between 1700 and 825 4 cm^{-1} contain the functional groups of interest.

Results and Discussion

The Fourier Transform Infrared Spectroscopy (FT-IR spectroscopy) is an effective method to describe the structural changes occurring during treatment. Figure 48 shows the FT-IR spectrum of the precursor material (Hardwood Lignin). The collected FT-IR data matched previous research found in publications of Boeriu et al. and Faix et al. [4-21, 4-22]. The results of the FT-IR spectroscopy are summarized in Table 11.

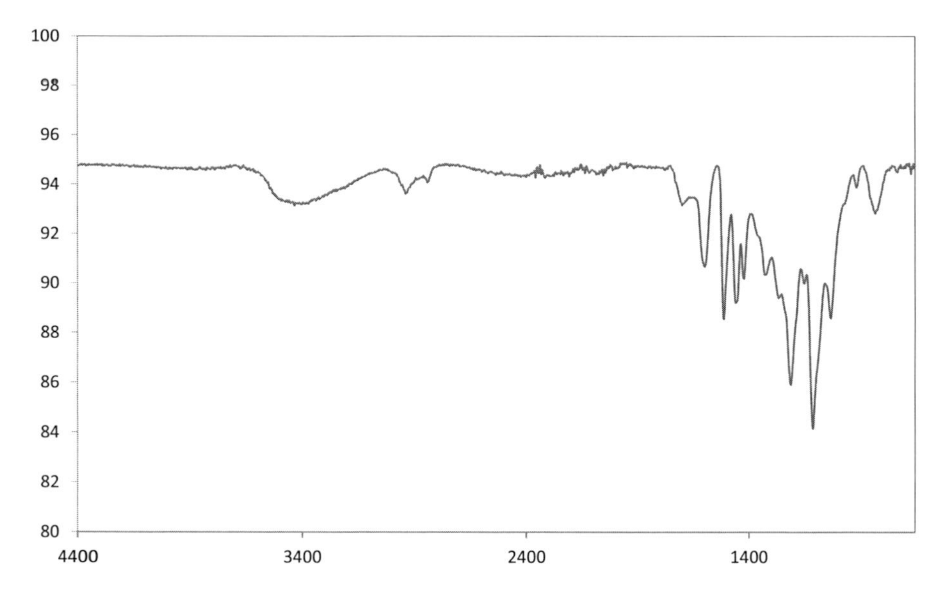

Figure 57: FT-IR Spectrum Hardwood-Lignin

Table 11: IR-Bands

Band	Intensity	Chemical Group
3400 – 3460 cm^{-1}	Bright	O-H stretching in aromatic and aliphatic structures
2936 – 2840 cm^{-1}	Medium	C-H stretching of methoxy groups, methyl and methylene groups in side chains
1701 – 1705 cm^{-1}	Medium	C-O stretching of unconjugated carbonyl and carboxyl groups
1598 – 1600 cm^{-1} 1513 – 1514 cm^{-1} 1423 - -1424 cm^{-1}	Strong	typical aromatic skeleton vibrations
1459 – 1462 cm^{-1}	Strong	aromatic ring vibration
1365 – 1368 cm^{-1}	Weak	O-H stretching of phenolic OH-groups
1326 – 1327 1267 – 1270	Medium	C=O stretching at the ring
1212 – 1214	Strong	the C-C, C-O and C=O stretching
1151 – 1112 1031 – 1130	Medium	C-H stretching in plane deformation of the ring
829 – 831	Bright	aromatic C-H out-of-plane deformation vibrations

The Hardwood Lignin consists of a composition of syringyl and guaiacyl units (Figure 58). For both aromatic units the typical aromatic skeleton vibrations were found at 1600, 1514 and 1424 cm^{-1} and the aromatic ring vibration at 1462 cm^{-1}.

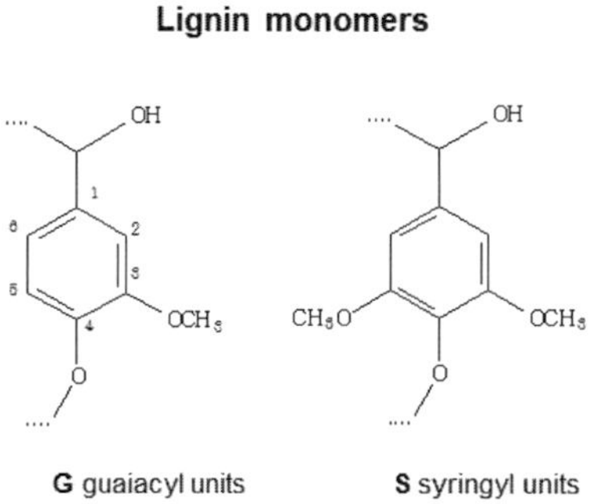

Lignin monomers

G guaiacyl units **S** syringyl units

Figure 58: Hardwood Lignin monomers

The region in the FT-IR Spectrum below 1400 cm^{-1} is more difficult to analyze. In this region combinations of different vibration states can influence the spectrum.

The band at 1326 cm^{-1} and 1270 cm^{-1} are the rings breathing with C=O stretching and characteristic for S and G units, respectively (Compare Figure 58). At band 1113 cm^{-1} the C-H group of the S unit has the highest intensity of vibration due to in plane deformation. The next highest intensity of vibration for both S and G units is a combination of C-C, C-O and C=O stretching at the band at 1214 cm^{-1}. The C-H group of the guaiacyl units has an in plane deformation vibration at the band at 1031 cm^{-1}. The smaller band at 831 cm^{-1} is observed and characteristic for the aromatic C-H out-of-plane deformation vibrations of the S units (Figure 58).

The results of the FT-IR Spectroscopy confirm the structures established with the Nuclear Magnetic Resonance Spectroscopy, Mass Spectroscopy and Elementary Analysis. FT-IR Spectroscopy will also be a useful tool when defining the reaction mechanism during conversion of Hardwood Lignin to carbon fiber. Chapter 4.3 will put all the results together to define the chemical structure of Hardwood Lignin powder as a precursor for carbon fiber production.

4.3 *Chemical Structure of Hardwood Lignin for carbon fiber production*

To define the structure of Hardwood lignin it was necessary to consider the results from the Elementary Analysis (EA), Mass Spectroscopy (MS), Nuclear Magnetic Resonance Spectroscopy (NMR), and Fourier Transform Infrared Spectroscopy (FT-IR). Using the results of the EA of Hardwood Lignin it was confirmed that the basic structure is the C_9 Unit (Figure59).

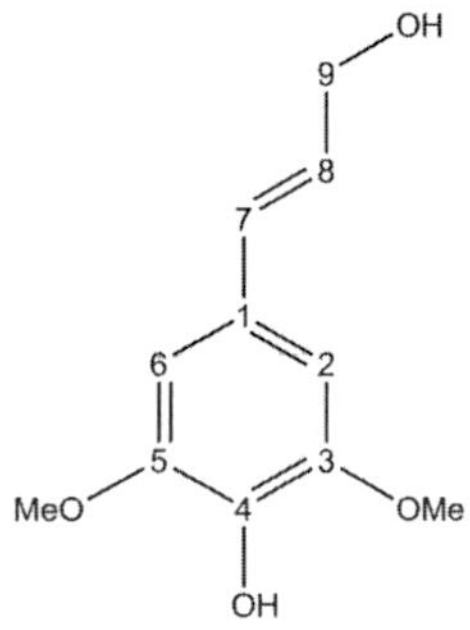

Figure 59: C_9 unit calculated with the help of Elementary Analysis

Mass spectrometer allowed quantitation of atoms or molecules and provides structural information by the identification of distinctive fragmentation patterns. Using Nuclear Magnetic Resonance (NMR) Spectroscopy the identification of the monomers and side chains was possible. To validate these results a Fourier Transform Infrared Spectroscopy (FT-IR) was performed. The combination of MS, NMR, and FT-IR results provided the structure for the Hardwood Lignin monomers (Figure 60) and the Hardwood Lignin side-chains (Figure 61).

Since the HSQC measurements were successfully used for the detection of the structure of Hardwood Lignin, quantitative measurements were done by performing the same test in a pulsing mode. Using the results it is visible that there is a relationship between the monomers G and S, which is approximately 1: 2.2. That means there are 2.2 times more syringyl units than guaiacyl units (Figure 59).

The same calculations were made for the side chains, where the bottom part is always a syringyl unit: relation between A, B and R is 1: 1.8 : 2.8 (Table 12 and 13).

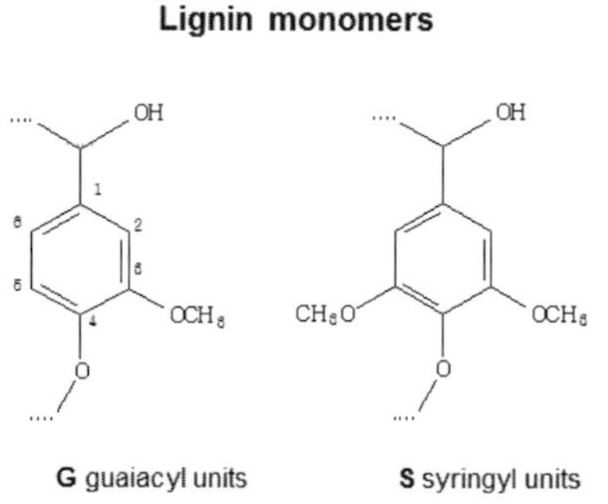

Lignin monomers

G guaiacyl units **S** syringyl units

Figure 60: Monomers found in Hardwood Lignin

Table 12: Relation between monomers in Hardwood Lignin

	Guaiacyl	Syringyl
Relation	1	2.2

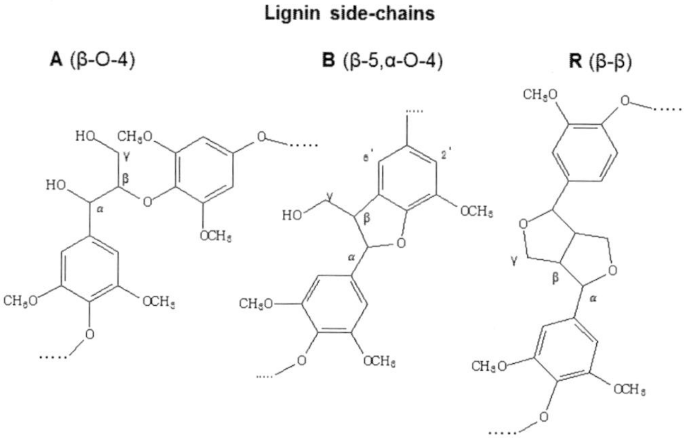

Lignin side-chains

A (β-O-4) **B** (β-5,α-O-4) **R** (β-β)

Figure 61: Lignin side-chains found in Hardwood Lignin

Table 13: Relation between side chains in Hardwood Lignin

	A	B	R
Relation	1	1.8	2.8

On the basis of these results, a chemical structural formula for the used Hardwood Lignin can be proposed (Figure 61).

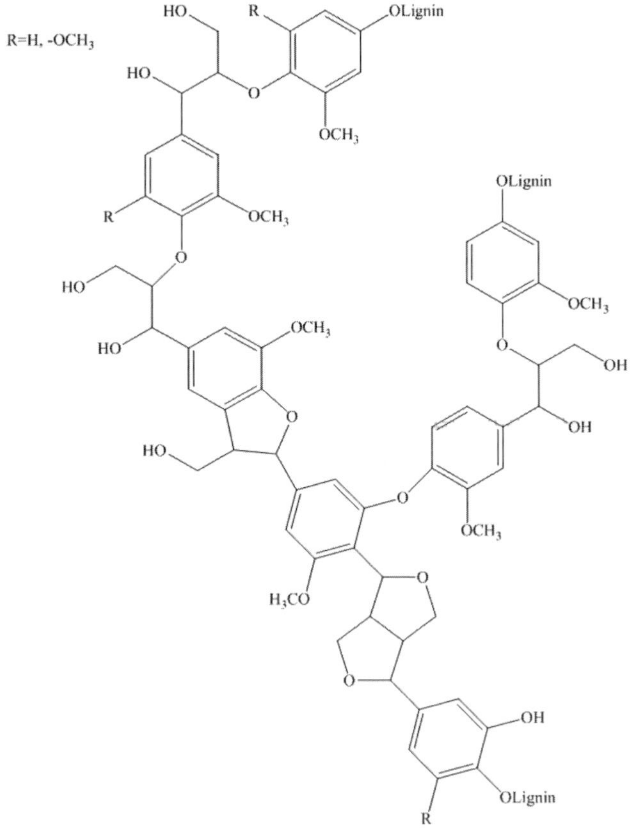

Figure 62: Chemical structure of the used Hardwood Lignin

4.4 References

[4-1] Ehrenstein, G. W. ; Riedel, G. ; Trawiel, P.:
Praxis der thermischen Analyse von Kunststoffen. Hanser Verlag, 2003

[4-2] Brebu, Mihai ; Vasile, Cornelia: Thermal degradation of lignin - a review.
Cellulose Chemistry & Technology 2010; 44: 353 – 361

[4-3] Ehrenstein, G. W. ; Riedel, Gabriela ; Trawiel, Pia:
Praxis der thermischen Analyse von Kunststoffen. Hanser Verlag, 2003

[4-4] El-Sabbagh, A: Effect of coupling agent on natural fibre in natural fibre/
polypropylene composites on mechanical and thermal behaviour.
Composites Part Engineering 2014, 57: 126–135

[4-5] In: Glasser, Wolfgang G.:
Classification of lignin according to chemical and molecular structure.
AMER CHEMICAL SOC 1155 SIXTEENTH ST NW, WASHINGTON, DC
20036 USA, 2000; 742: 216–238

[4-6] Bouajila, J ; Dole, P ; Joly, C ; Limare, A: Some laws of a lignin plasticization.
Journal of applied polymer science 2006; 102: 1445–1451

[4-7] N. E. Mansouri , Joan Salvado: Analytical methods for determining unctional
groups in various technical lignins.
Industrial Crops and Products 2007; 26: 116–124

[4-8] J. Liebig: Über einen neuen Apparat zur Analyse organischer Körper, und die
Zusammensetzung einiger organischer Substanzen.
Annalen der Physik 1831; 21: 1 – 47

[4-9] A. G. Marshall, C. L. Hendrickson: High-Resolution Mass Spectrometers.
Annu. Rev. Anal. Chem. 2008; 19: 1–21

[4-10] M. S. B. Munson, F. H. Field: Chemical Ionization Mass Spectrometry -
General Introduction.
Journal of the American chemical society: 1966; 88: 2621-2630

[4-11] Aue, W., E. Bartholdi, R.R. Ernst, Two - dimensional spectroscopy.
Application to nuclear magnetic resonance. The Journal of Chemical Physics,
1976. 64: p. 2229.

[4-12] Gomathi, L., Elucidation of secondary structures of peptides using high resolution NMR. Current Science, 1996. 71(7): p. 553.

[4-13] Ames, J.B., Hamasaki, N., Molchanova, T., Structure and calcium-binding studies of a recoverin mutant (E85Q) in an allosteric intermediate state. Biochemistry, 2002. 41(18): p. 5776. DOI: 10.1021/bi012153k

[4-14] R. R. Ernst, W. A. Anderson: Application of fourier transform spectroscopy to magnetic resonance. Review of Scientific Instruments 1966; 37: 93-201

[4-15] W. P. Aue, E. Bartholdi, R. R. Ernst: Two-dimensional spectroscopy. Application to nuclear magnetic resonance. The Journal of Chemical Physics 1976; 64: 2229-2246

[4-16] J.-L. Wen, S.-L Sun, B.-L XueR-C. Sun. Recent Advances in Characterization of Lignin Polymer by Solution Nuclear Magnetic Resonance (NMR) Methodology. Materials 2013; 6: 359-391.

[4-17] J. Ralph, L.L. Landucci. NMR of Lignins. In Lignin and Lignans: Advances in Chemistry; C. Heitner, D.R. Dimmel, J.A. Schmidt, Eds.; CRC Press: Boca Raton, FL, USA, 2010: 137-234.

[4-18] M. Foston, G.A. Nunnery, X. Meng, Q. Sun. NMR a critical Tool to Study the production of carbon fiber from lignin, Carbon. 2013; 52: 65-73

[4-19] Vollhardt, K. Peter C., and N. Schore. Organic chemistry structure and function.
New York: W.H. Freeman, 2009. pp. 468–503

[4-20] Faix, O., 1992. Fourier transformed infrared spectroscopy. In: Lin, S.Y., Dence, C.W. (Eds.), Methods in Lignin Chemistry. Springer-Verlag, Berlin-Heidelberg, pp. 458–464.

[4-21] Boeriu, CG, Bravo, D, Gosselink, RJA, van Dam, JEG. Charakterisation of structure-dependent functional properties of lignin with infrared spectroscopy. Ind. Crops and Prod. 2004; 20: 205-218.

[4-22] O. Faix, O. Beinhoff, FTIR spectra of milled wood lignins and lignin polymer models with enhanced resolution obtained by deconvolution, J. Wood Chem. Technol. 1988; 8(4): 505-522.

5 Development of a Lignin Based Carbon Fiber

The following section will show the process of making a lignin based carbon fiber in a semi-production scale [5-1, 5-2].

The conventional carbon fiber production process of a PAN based carbon fiber contains fiber spinning, fiber stabilization, and fiber carbonization as the major steps (compare chapter 3). For a future scale up of the lignin based carbon fiber production it is necessary to create a process which is comparable to the PAN based carbon fiber production process. This will also help to lower the capital cost for future production facilities, because conventional conversion equipment can be used for lignin carbon fiber production.

Compared to the conventional PAN based carbon fiber production process, two steps were added to prepare the fiber spinning process. These steps are washing and drying of lignin powder as the first step, and pelletizing of the lignin powder as the second step.

This leads to five main steps, which illustrate the use of Hardwood Lignin as a precursor for carbon fiber production:

- Washing and drying of lignin powder
- Pelletizing of the lignin powder
- Melt spinning of the lignin fiber
- Oxidation of the fiber
- Carbonization of the fiber

The following sections will describe the five steps in more detail and will show the production process, as performed at Oak Ridge National Laboratory in Tennessee, USA.

The lignin based carbon fiber project was done with the approval of the U.S. Department of Energy under the Cooperative Research and Development Agreement (CRADA) NFE-12-03992. For this reason, the details in this thesis are limited.

5.1 Lignin Powder

The isolation process of Hardwood Lignin is described in chapter 3.5. A detailed analysis of the properties and the chemical characterization of the Hardwood Lignin precursor powder can be found in chapter 4.1 and 4.2.

At the paper mill, lignin is dissolved away from the cellulose in the black liquor solution. To isolate the lignin it must be precipitated out of the black liquor solution. Before the lignin powder can be pelletized, it must go through a washing and drying process. An example of the Hardwood Lignin powder, how it was received from the lignin isolation process, is shown in Figure 63 [5-1, 5-2].

Figure 63: Hardwood Lignin powder

5.2 Compounding and Pelletizing

The Hardwood Lignin powder can not be used for fiber spinning. The main reason is the bad flow ability of the Hardwood Lignin powder. The reason for bad flow ability are high components of moisture and volatiles, which can be detected in the hydrophilic Hardwood Lignin powder. To improve the flow ability and reduce the moisture and the volatiles compounds, the Hardwood Lignin powder must be pelletized.

The twin screw compounding extrusion process significantly reduced the moisture and volatiles compounds. The reduction of the moisture is done by the conventional vaporizing extrusion and the addition of vacuum port. The conventional method of removing moisture during an extrusion did not meet the goals to spin a lignin fiber. To further reduce moisture and volatiles compounds, the addition of a vacuum was necessary (Figure 64). This concept was proved with a 27 mm extrusion machine. A Hardwood Lignin powder throughput of 45 kg/hour was demonstrated.

For making carbon fiber mainstream, it is necessary to demonstrate large scale production of the pelletizing process. This was done with a 53 mm diameter extrusion machine, which was equipped with a hot die face cutter. 1000 kg of Hardwood Lignin was successful pelletized for subsequent melt spinning into Hardwood Lignin precursor fiber [5-1, 5-2]. The in laboratory scale pelletized Hardwood Lignin is shown in Figure 65.

Figure 64 shows the 53 mm diameter extrusion machine with following dimensions: Centerline Height 1090 mm; Length 5410 mm; Width 1030 mm; Height 1470 mm and Weight 5440 kg.

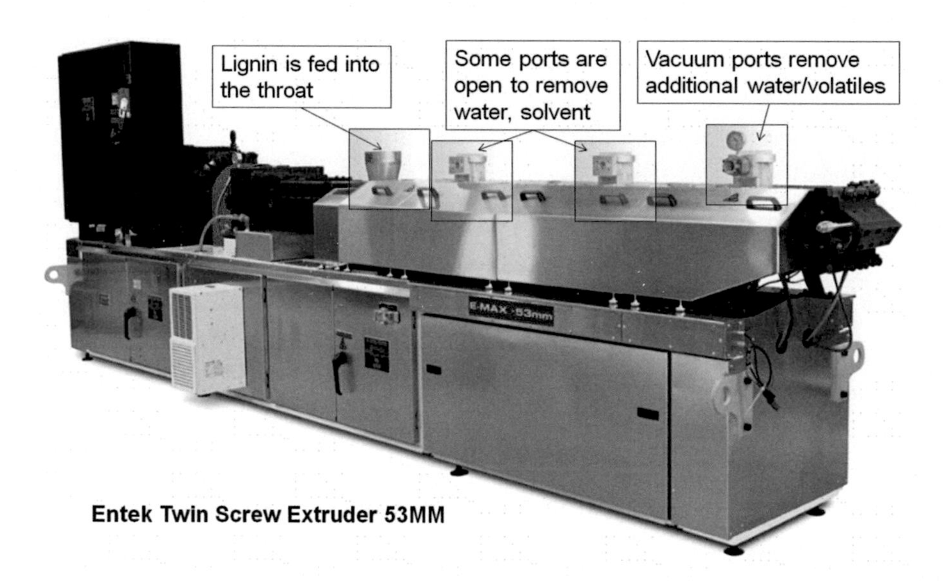

Figure 64: 53 mm diameter extrusion machine used for pelletizing lignin

NMR and FT-IR analysis was performed on the Hardwood Lignin pellets to understand, what influences the compounding and pelletizing process had on the chemical structure of the material. The results of the characterization and the differences between the chemical structures can be found in chapter 6.

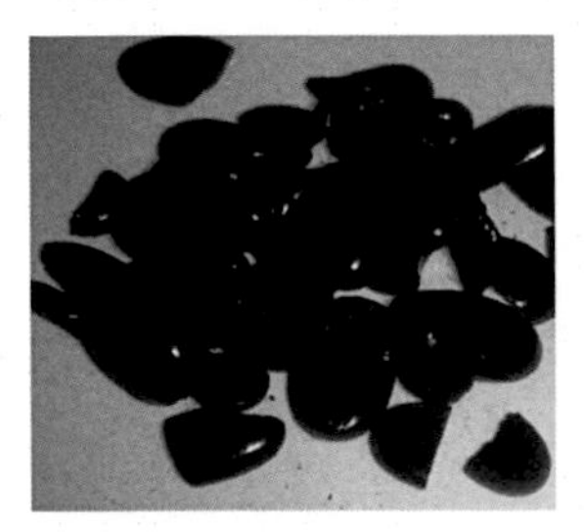

Figure 65: Pellets made from Hardwood Lignin

5.3 Precursor Fiber Production

The most important step during production of a Hardwood Lignin based carbon fiber is the spinning process. The quality of the precursor fiber defines the quality of the carbon fiber after conversion.

Using the thermal properties of the Hardwood Lignin (Chapter 4.1), single fiber spinning experiments were performed to identify the parameter necessary for the melt spinning process (temperature profile, kind and amount of plasticizer, spinning speed etc.). After several single fibers spinning tests were completed, a scale up with a melt blow spinning system was made.

Using a laboratory scale machine, “melt blown” spinning was tested for producing a Hardwood Lignin fiber web. The filament diameter ranged from 10 to 20 µm. At rates approaching 15 kg/h, the fibers were spun into a web approximately 60cm wide with a areal density of 230 g/m^2. Figure 66 illustrates the precursor fiber production using the melt blow process [5-3].

This is the first presented process which shows it is possible to produce Hardwood Lignin fibers in a Semi-production scale.

Figure 66: Melt blow process for producing lignin fiber

By performing more the 100 single fiber spinning experiments, minimum requirements of the lignin quality were established. These parameters ensure that the lignin is capable of being melt spun [5-3]:

- >99% lignin
- <500 ppm residual carbohydrates
- <5 wt% volatiles
- <1000 ppm ash
- <500 ppm non-melting particles larger than 1 µm diameter

The definition of Hardwood Lignin specifications for carbon fiber production is also completely new (no specification like this available in literature). This will help to define a profile of properties for lignin producers, by defining how and which lignin is useful for carbon fiber production.

A detailed characterization of the properties and the chemical structure of the lignin based precursor fiber can be found in chapter 7 and 8.

5.4 Fiber Stabilization

The fiber stabilization is similar to the PAN based carbon fiber production process the most critical step during conversion. The oxidation process must be handled very carefully to prevent any damage of the fiber. If the oxidation conditions are too harsh, a complete oxidation of the fiber is possible. The result is a fire in the oxidation oven and destroying of the fiber by producing CO_2 and ash. To prevent such issues, the stabilization time was extended and a gradual temperature ramp was chosen.

Stabilization time was and remains a significant challenge. For the Hardwood Lignin, in laboratory scale, the stabilization time requires several days. It was possible to show that a successful stabilization is possible at temperatures between 200 and 300 °C by using a gradual temperature ramp (Figure 67) [5-1].

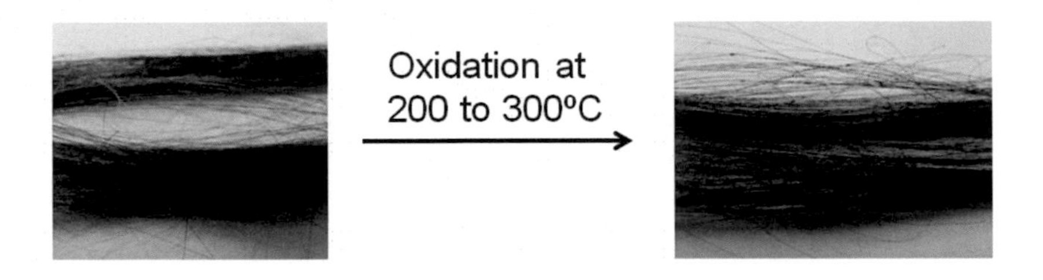

Figure 67: Stabilizing of lignin fiber

Stabilization must be accelerated to achieve acceptable process economics. A simple tuning of the thermal profile, in semi-production scale, reduced the residence time from ~150 hours to ~ 100 hours. That means a time reduction of more than 33% was achieved.

A comparison with the conventional stabilization process of PAN based carbon fiber (stabilization time of 90 minutes) shows the disadvantages of a batch process in laboratory scale compared to an industrial optimized continuous process. Future research has to show the process ability of lignin based carbon fiber in a continuous process.

In laboratory scale the stabilized fibers were produced using a batch thermal treatment process. The fibers were placed in a large (> 5.5m³) oven (Figure 68), using a stabilization time of ~100-hr. In one batch 75kg of lignin fibers were stabilized for further processing. This is the first reported stabilization of lignin fibers at a scale exceeding 1 kg and shows the possibility to scale up the stabilization in industrial scale.

The material darkens in color from brown to black (Figure 67), as it is heat treated in the oxidation oven (Figure 68) [5-2, 5-3].

Figure 68: Batch stabilization of Hardwood Lignin fibers

Additional reductions of the stabilization time are possible by future optimization of the temperature ramp and the concentration of oxygen in the stabilization oven. Also a transfer of the results from a batch process to a continuous process will help to improve the economics of the stabilization process. It should be possible to reduce the stabilization time, in an industrial process, below 3h (compare conventional process using PAN precursor – Chapter 2.2).

A detailed characterization of the properties and the chemical structure of the stabilized lignin fiber can be found in chapter 7 and 8.

5.5 Fiber Carbonization

The fiber carbonization is the last step in the production of lignin based carbon fiber. The stabilized material is then carbonized in a furnace under nitrogen atmosphere. The carbonization process can occur at temperatures ranging from 500 to 1500 °C for 5 to 10 min. Approximately 65% of the material is vaporized during carbonization, with gasses exhausted through an incineration system. In the laboratory, the stabilized fibers were heat-treated to produce ~ 25 kg of lignin based carbon fibers. This is the first reported carbonization of lignin fibers at a scale exceeding 1 kg and shows the possibility to scale up the stabilization in industrial scale.

The remaining material is nearly 100 percent pure carbon – a lignin-based carbon fiber (Figure 69) [5-1]. A detailed characterization of the properties and the chemical structure of the produced carbon fiber can be found in chapter 7 and 8.

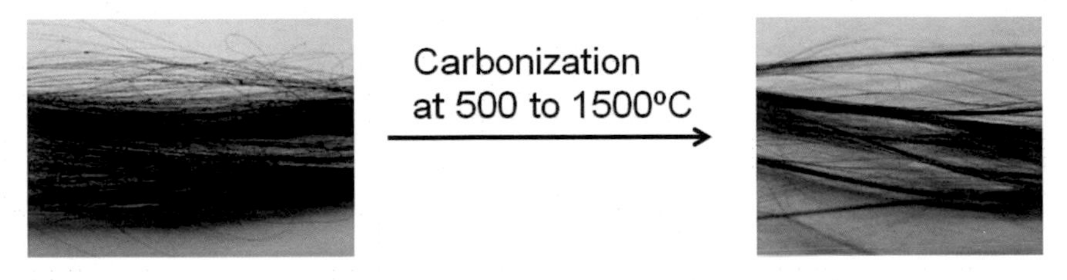

Figure 69: Carbonizing of stabilized lignin fiber

Through laboratory testing, a set of necessary process and materials parameters for producing lignin based carbon fiber were established. These parameters were used in a semi-production scale process. In this semi-production scale 1000 kg of lignin pellets were produced. From these pellets, the lignin precursor fiber was successfully spun, stabilized, and carbonized to produce the largest batch of lignin based carbon fiber ever produced. With 75 kg of produced carbon fiber, it has now been proven that lignin is viable precursor for carbon fiber mass production.

5.6 References

[5-1] H. Mainka, O. Täger, O. Stoll, E. Körner, A. S. Herrmann: Alternative Precursors for Sustainable and Cost-Effective Carbon Fibers usable within the Automotive Industry.
Society of Plastics Engineers (Automobile Division) – Automotive Composites Conference & Exhibition 2013, Novi, Mich. USA

[5-2] H. Mainka, E. Körner, O. Täger, A. Plath L. Hilfert, S. Busse, F. Edelmann, A.S. Herrmann: Lignin — An Alternative Precursor for Sustainable & Cost-Effective Automotive Carbon Fiber.

Society of Plastics Engineers (Automobile Division) – Automotive Composites Conference & Exhibition 2014, Novi, Mich. USA

[5-3] C. Eberle, T. Albers, C. Chen ,D. Webb Commercialization of New Carbon Fiber Materials Based on Sustainable Resources for Energy Applications (ORNL/TM-2013/54) Published: March 2013

6 Major reactions during conversion of lignin

For qualifying lignin based carbon fiber for automotive mass production a detailed characterization of this new material is necessary. Therefore nuclear magnetic resonance spectroscopy and Fourier transform infrared spectroscopy is used. Using the results of these experiments the major reactions during conversion of lignin to carbon fiber are proposed.

6.1 Sample Characterization

For qualifying lignin based carbon fiber for automotive mass production a detailed characterization of this new material is necessary. Therefore nuclear magnetic resonance spectroscopy, Fourier transform infrared spectroscopy and elementary analysis were used.

The HSQC experiments were carried out with a Bruker AVANCE 600 NMR spectrometer operating at 600.13 MHz and 298 K. The spectrometer was fitted with a 5 mm TBI-1H-13C/15N/2H probe head with z-gradients. The 1H and 13C chemical shifts were determined relative to internal DMSO-d6 and were given in parts per million downfield to TMS.

HSQC were separately measured for the aliphatic and for the aromatic part with spectral width in F2 of 1.3 kHz and 2.7 kHz, respectively. For all experiments the spectral width was in F1 8 kHz. The HSQC were detected using 2 K data points in F2 and 256 experiments of 128 scans in F1, relaxations delay was 1.5 s and pulse width for 1H was 9.3 µs and for 13C was 13.0 µs.

The 31P NMR spectra were obtained by using inverse gated decoupling on a Bruker Avance 400 NMR spectrometer, operating at 161.9 MHz and using a 5 mm BBOF probe head with z-gradients. The external standard was 85% H3PO4. The lignin samples were dissolved in 0.75 ml of a mixture of deuterated pyridine and chloroform in ratio of 3:2 as solvent with 0.47 mol/l cyclohexanol as internal standard (145 ppm). The samples were in situ derivatizated in the NMR tube with 0.1 ml 2-chloro-4,4,5,5-tertamethyl-1,2,3-dioxaphospholane (TMDP). The inverse gated decoupling se-

quence was used with 25 s relaxation delay and 128 scans were collected.
The 13C CP/MAS spectra were obtained on a Bruker Avance 300 NMR spectrometer, operating at 75.47 MHz. The spectrometer was equipped with 4 mm CP/MAS probehead BBO/ML4. The samples were measured in 4 mm ZrO2 rotors MAS rotors. The proton 90° pulse was set to 2.55 µs and the decoupling strength during acquisition was 54 kHz. Following conditions were applied: 5 kHz spinning speed, recycle delay 3 s, contact time 1.74 ms. Chemical shifts are quoted in ppm from TMS. The sample were pulverized and mixed with equal weight of hydrated magnesium silicate to reduce the macro-current generation under radio frequency excitation.
Infrared spectra were recorded on a Perkin Elmer FT/IR 2000. The samples were measured using an ATR unit. The spectral resolution was 4 cm-1. The amount of sample used in IR could not be kept constant. Therefore only the relative ratios of the interesting functional groups were compared.
The investigations of the contained elements in the used Hardwood Lignin for carbon fiber production are done with the Vario ELcube CHNS from Elementar. The sulfur-carbon-analyzer CS230 and the CHN-analyzer from LECO were used. All measurements were repeated at least three times and the average of the results is shown.

6.2 Results

6.2.1 Nuclear Magnetic Resonance Spectroscopy

^{1}H and ^{13}C NMR spectroscopy

The evolution during the treatment was investigated by elementary analysis (EA), Infrared Spectroscopy, solution NMR and solid state NMR (CP-MAS). The complex natural structure and the associated limited solubility are important factors for the applicability of the analytical methods. Figure70 summarizes the chemical structures of the monolignols, the inter unit linkages and the functional groups of Hardwood Lignin. The NMR spectroscopy is a powerful technique for the characterization of lignin structure. Capanema et al [22, 23] have used a combination of correlation 2D NMR methods and a quantitative 1D 13C NMR technique for a comprehensive approach of lignin structures. In this paper we investigated exclusively non-acetylated lignin samples to avoid the loss of components.

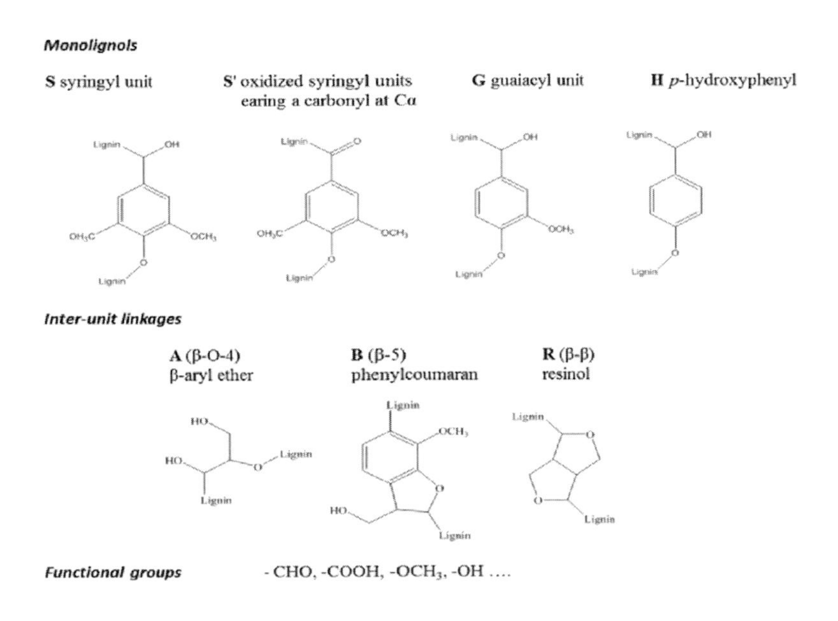

Figure 70: Main chemical structures in lignin

At first 2D correlation NMR spectra were discussed to determine structural modification before quantification. The 2D NMR spectroscopy was used to identify the primary structures of the investigated lignin derivatives. The assignments of ^{1}H and ^{13}C chemical shifts are based on the HSQC (Heteronuclear Single-Quantum Correlation) and HMBC (Heteronuclear Multiple Bond Correlation). The spectra were divided in three parts – the aromatic, side-chain and aliphatic. The signals in aliphatic region are not sensitive for structural modifications. For the structure elucidation the aromatic and side-chain regions are of particular interest to identify the lignin monomers and inter-unit linkages. The HSQC and HMBC spectra for the Hardwood Lignin and pellets were presented in Figure 71. For every sample the aromatic and side-chain part were separately scanned. The cross-peaks were assigned by comparing the literature [6-1, 6-2, 6-3] and listed in Table 14 and Table 15.

Table 15 shows the assignments of ^{13}C-^{1}H correlation signals of the side chain region in the HSQC spectra of Hardwood Lignin (HW) and pellets (P), which were made from this lignin. Table 15 shows the assignments in the aromatic region.

Table 14: Assignment of ^{13}C-^{1}H correlation signals of side-chain region in the HSQC spectra of HW and P

	HW		**P**		
Label	^{13}C in ppm	^{1}H in ppm	^{13}C in ppm	^{1}H in ppm	**Assignment**
OMe	n.o.	n.o.	51.5	3.56	**O-CH_3**
R_β	54.0	3.05	54.1	3.06	C_β in **R**
B_β	54.3	2.83	54.3	2.83	C_β in **B**
O-CH_3	56.3	3.74	56.4	3.75	**O-CH_3**
O-CH_3	n.o	n.o.	59.2	3.24	O-CH_3 in **A'**
A_γ	60.2	3.57	60.2	3.57	C_γ in **A**
A_γ	61.0	4.12	n.o.	n.o.	C_γ in **A**
A'_γ	-	-	60.2	4.01	C_γ in **A'**
B_γ	64.1	3.27	64.6	3.32	C_γ in **B**
R_γ	70.4	4.08 / 3.73	70.8	4.09/3.76	C_γ in **R**
R_γ	71.4	4.16 / 3.75	71.4	4.16/3.78	C_γ in **R**
A_α	71.0	4.75-4.85	72.3	4.88	C_α in **A** linked to **G**
A_α	72.0	4.75-4.85	72.3	4.88	C_α in **A** linked to **S**
A_β	81.5	4.75	81.9	4.75	C_β in **A** linked to **G**
R_α	85.7	4.60	85.6	4.65	C_α in **R**
A_β	85.7	4.60	85.6	4.65	C_β in **A** linked to **S**
B_α	87.4	4.30	87.3	4.32	C_α in **B**

Table 15: Assignment of ^{13}C-^{1}H correlation signals of aromatic region in the HSQC spectra of HW and P

	HW		**P**		
Label	^{13}C in ppm	^{1}H in ppm	^{13}C in ppm	^{1}H in ppm	**Assignment**
$S_{2/6}$	n.o.	n.o.	101.2	6.89	$C_{2/6}$-$H_{2/6}$ in **S**
$S_{2/6}$	103.4	6.58	103.2	6.58	$C_{2/6}$-$H_{2/6}$ in **S** linked to **A**
$S_{2/6}$	103.7	6.59	103.7	6.60	$C_{2/6}$-$H_{2/6}$ in **S** linked to **R**
$S_{2/6}$	103.7	6.61	103.7	6.63	$C_{2/6}$-$H_{2/6}$ in **S** linked to **B**
$S_{2/6}$	103.7	6.83	103.9	6.83	$C_{2/6}$-$H_{2/6}$ in **S**
$S_{2/6}$	105.3	6.98	n.o.	n.o.	$C_{6'}$-$H_{6'}$ in **B**
$S_{2/6}$	105.3	7.13	n.o.	n.o.	$C_{2/6}$-$H_{2/6}$ in **S**
	n.o.	n.o.	105.5	6.85	
	n.o.	n.o.	105.5	6.87	
$S_{2/6}$	105.3	6.69	105.6	6.85	$C_{2/6}$-$H_{2/6}$ in **B**
$S'_{2/6}$	106.6	7.22	106.8	7.05	$C_{2/6}$-$H_{2/6}$ in oxidized **S**
well-suiteS'$_{2/6}$	107.5	7.19	107.6	7.20	$C_{2/6}$-$H_{2/6}$ in oxidized **S**

$S'_{2/6}$	108.3	7.16	n.o.	n.o.	$C_{2/6}$-$H_{2/6}$ in oxidized **S**
G_2	109.6	7.12	109.7	7.14	C_2-H_2 in **G**
	n.o.	n.o.	110.9	7.52	
	n.o.	n.o.	113.2	6.76	
G_5	115.6	6.6-6.6	115.6	6.7-6.9	C_5-H_5 in **G**
G_5	115.2	6.6-6.8	115.8	6.7-6.9	C_5-H_5 in **G**
G_6	119.6	6.93	119.9	6.93	C_6-H_6 in **G**
	n.o.	n.o.	118.8	6.75	
	n.o.	n.o.	123.3	7.52	
$H_{2/6}$	125.6	6.93			
$H_{2/6}$	126.2	6.93	125.9	6.95	$C_{2/6}$-$H_{2/6}$ in **H**
$H_{2/6}$	126.4	6.82	n.o.	n.o.	$C_{2/6}$-$H_{2/6}$ in **H**
	n.o.	n.o.	127.9	5.29	C=C, olef.
	n.o.	n.o.	129.9	5.31	C=C, olef.
S_1, G_1	133.6-135.4	-	135.4-137.6	-	C_1 in S_1, G_1
S'_1	140.0-141.8	-	142.3	-	C_1 **in S'_1**
S_4, $S_{3/5}$	147.1-148.2	-	148.3		C_4, $C_{3/5}$ in etherified S
G_3, G_4	128.1/129.5	-	128.6/129.2	-	C_3, C_4 in etherified **G**
C=O	n.o.	-	168.3		C=O in acetates
C=O	190.8 – 196.3	-	191.5	-	α-CO / γ-CHO

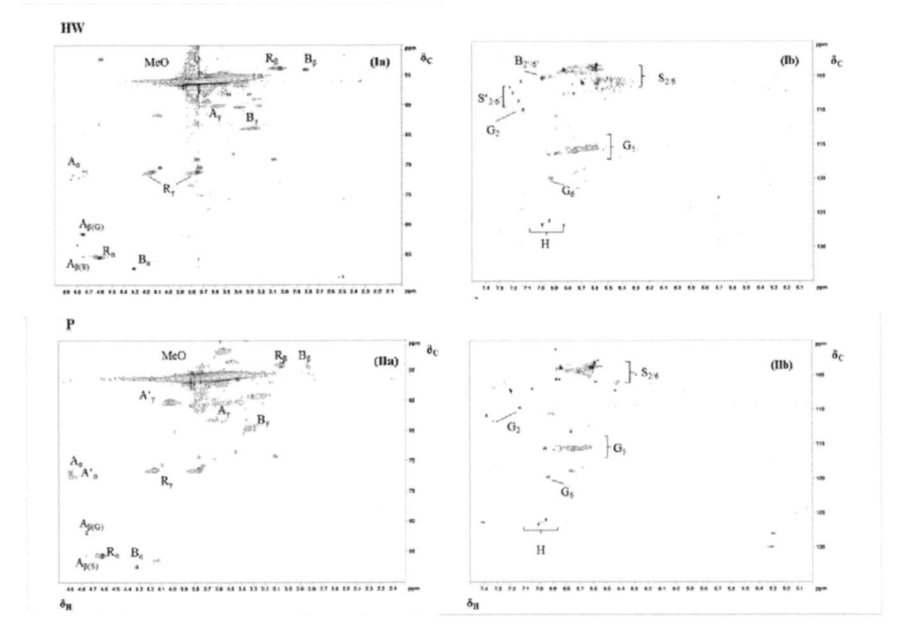

Figure 71: HSQC spectra for the side-chain region (Ia and IIa) and the aromatic region (Ib and IIB) for HW and P, respectively

The side-chain region is ranging from 50 - 90 ppm for ^{13}C and from 2.0 – 5.0 ppm for

^{1}H chemical shifts. Besides the main signal δC/δH 56.3/3.74 ppm for the methoxy group occur the signals for substructure A for Cα-Hα in the range at δC/δH 71.0-72.0/4.75-4.85 ppm and for Cβ-Hβ at 81.5/4.75 ppm and 87.7/4.60 ppm linked to G and S, respectively. For Cγ-Hγ the cross peaks were observed at δC/δH 60.2/3.57 ppm. Furthermore strong signals were observed for resinol substructure R at δC/δH 54.0/3.05 ppm for Cβ-Hβ, 71.4/4.75 and 3.37 ppm for Cγ-Hγ and 87.7/4.60 ppm for Cα-Hα. Also phenylcoumaran B is a well-known inter-unit linkage in lignis and were found in the HW and P samples. The signals were detected at δC/δH 54.3/2.83 ppm for Cβ-Hβ, 64.1/3.27 ppm for Cγ-Hγ and 87.4/4.30 ppm for Cα-Hα. The investigated samples were produced from Hardwood Lignin and typically contain guaiacyl G and syringyl S units.

The syringyl S and guaiacyl G could be assigned in the aromatic region ranging from 100 – 140 ppm for ^{13}C and 5.0 -7.5 ppm for ^{1}H. The S units showed different correlations for C2/6-H2/6 δC/δH 103.4-105.3/6.58-6.87 ppm depend on the different inter-unit linkages at C4 and C1. The C2/6-H2/6 correlations in Cα-oxidized S units were found at δC/δH 106.6-108.3/7.12-7.22 ppm. The G units revealed dominant signals for the C2-H2 at δC/δH 109.6/7.12, for C5-H5 broad signal at δC/δH 115.2-115.5/6.6-6.8 ppm and C6-H6 at δC/δH 119.6/6.93 ppm. Signals for H were detected at δC/δH 126.2-126.5/6.82-6.93 ppm.

The HMBC spectra (Figure 72) are examples for the efficiency of this technique. These spectra allowed the assignment of quaternary carbons C1, C3/5, C4 in S units, the carboxyl groups in oxidized syringyl groups in S' and the C1, C3 and C4 in G units. Strong long range correlations for the aromatic protons in S and G units were observed between S/G2/6, S/G5 and the inter-unit linkages Cα in B and R, Cβ in A and B.

The side-chain HSQC for HW in comparison with P are relatively identical. For the aromatic HSQC spectra we found significant differences. The signals for Cα in S' disappeared (IIb), additional olefin resonances are observed at δC/δH 127.9/5.29 and 129.9/5.31. It is noteworthy that the intensity and the number of the aromatic signals in the HSQC spectrum for P is lower than in the HSQC for HW. The 2D and the ^{13}C spectra indicate that there are no sugar contaminants in the preparations.

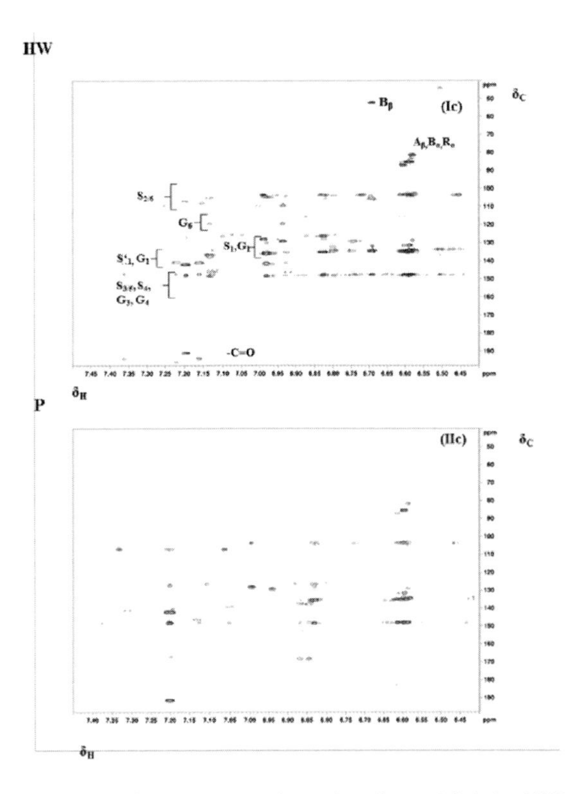

Figure 72: HMBC spectra for the aromatic region (Ic and IIc) for HW and P, respectively

^{31}P NMR spectroscopy

The ^{31}P NMR spectroscopy is an established method to quantitatively determine the amounts and distributions of the hydroxyl and phenolic groups in lignin and pretreated lignin [6-4, 6-5].

The starting material HW and the pretreated samples P, LF and OLF were analyzed followed as described in the literature [6-4, 6-5]. Cyclohexanol was used as internal standard. In the ^{31}P spectra are found typical chemical shifts ranges for aliphatic OH, syringyl and guaiacyl units, carboxylic OH and H units. The assignments of the ^{31}P chemical shifts are displayed in Figure 73. Table 16 summarizes the distribution of

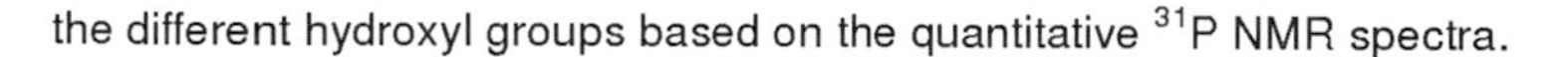
the different hydroxyl groups based on the quantitative ^{31}P NMR spectra.

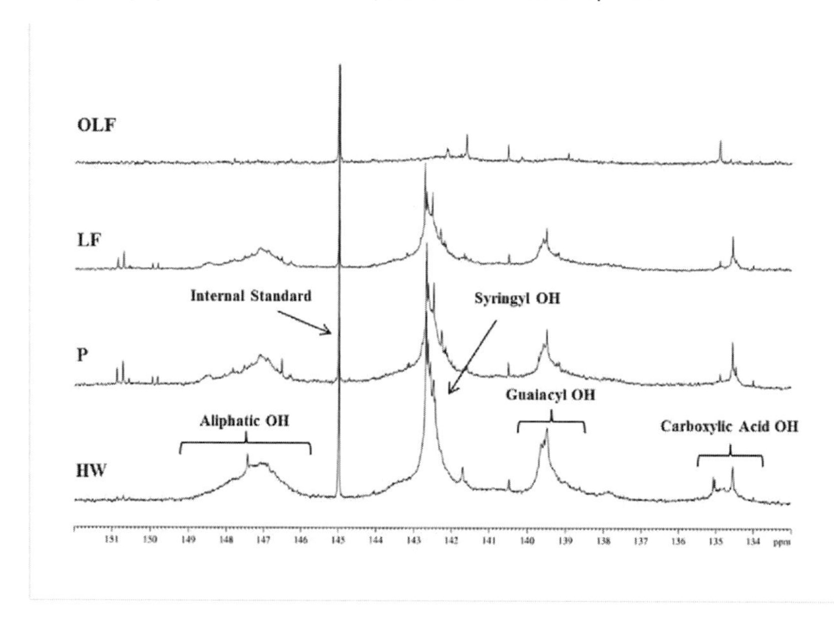

Figure 73: Quantitative ^{31}P NMR spectra and assignment of Hardwood Lignin (HW), pellets (P), lignin fiber (LF) and stabilized fiber (OLF) as a result of derivatization by TMDP

The concentration for the different hydroxyl groups was calculated relating to the peak area of derivatized hydroxyl groups of the internal standard. The NMR spectrum for the starting material HW shows a high amount of syringyl units (0.75 mmol*g-1) and aliphatic groups (0.53 mmol*g-1) and a lower amount for guaiacyl hydroxyl groups (0.24 mmol*g-1). For the carboxylic OH only minor amounts were found (0.03 mmol*g-1). After treatment the concentrations for all hydroxyl groups in P and LF decreased. This corresponds with the increasing cross linkages between the lignin macromolecules. The solubility of the OLF sample was very low and the discussion of the concentration is impossible.

Table 16: Distribution of hydroxyl and phenolic groups obtained by ^{31}P NMR analysis (mmol of OH/g of lignin)

Lignin sample	Aliphatic OH	S units	G units	Carboxyl	H units
HW	0.53	0.75	0.24	0.07	0.03
P	0.33	0.61	0.22	0.04	0.01
LF	0.35	0.53	0.15	0.05	0.01
OLF[a)]	n.o.[b)]	0.06	0.02	0.01	n.o.

^{13}C CP/ MAS spectroscopy

Figure 74 displays the ^{13}C CP/MAS NMR spectra of the complete process line. The assignments of the ^{13}C NMR signals are presented in Table 17 [6-8, 6-9, 6-12].

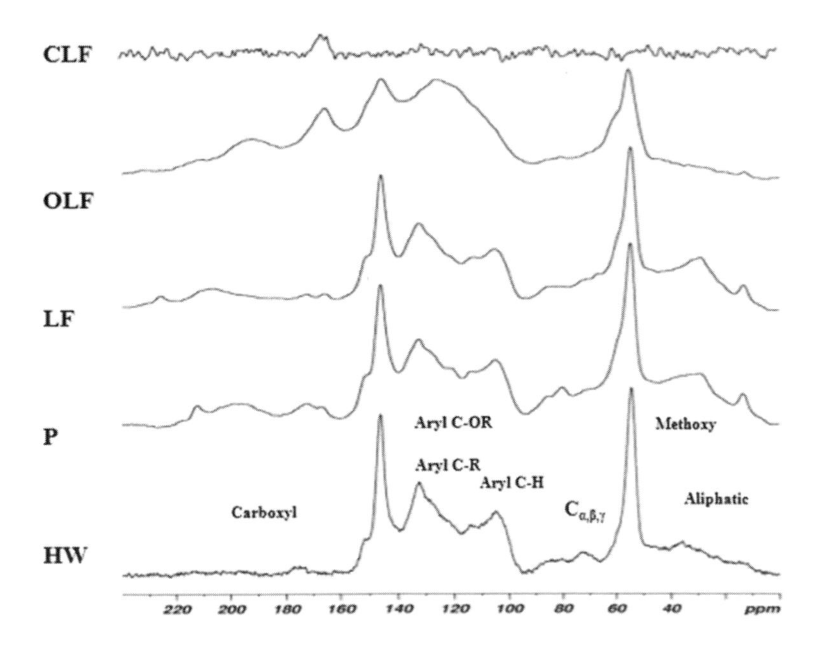

Figure 74: ^{13}C CP/MAS spectra with a MAS condition of 12 kHz, for OLF 30,000 scans and for CLF 70,000 scans were measured

Table 17: Assignment of ^{13}C CP/MAS spectra of HW and the treated samples P, LF, OLF and CLF

Label	HW	P and LF	OLF	CLF	Assignment
Aliphatic	13- 36	13 – 40	n.o.	n.o.	CH_3, CH_2, CH
Methoxy	50-54	50-54	42 – 62	n.o.	Aryl-OCH_3
$C_{\alpha,\beta,\gamma}$	73	71		n.o.	C_α
	81	80 - 84	n.o.		C_γ
	90	n.o.			C_α in B
Aryl C-H	103	104	100 – 138	n.o.	$S_{2/6}$
	109	n.o.			G_2
	117	113			$G_{5/6}$
Aryl C-R	132	132		n.o.	S_1, G_1 (no etherified)
Aryl C-OR	146-152	146-152	140 – 155	n.o.	$S_{3/5}$, S_4, G_3, G_4
Carboxyl	158 – 173	166 - 173	164	165	Aryl-COOR, anhydrides
	182	n.o.		n.o.	α-CO in S'
	201	190 - 203	190 – 200	n.o.	Aryl-CHO
	n.o.	212		n.o.	C=O (nonconjugated)

The starting material HW shows a typical CP MAS spectrum for Hardwood Lignin, carbonyl groups (200 -160 ppm), aryl (155 – 100 ppm), aliphatic side-chain (90 – 60 ppm), methoxy (50 - 54 ppm) and the aliphatic region (10 -30 ppm). The CP-MAS spectra for P and LF are very similar. In comparison with HW they show significant changes in the range of 160 – 225 ppm. The occurrence of ketones and/or aldehydes (190 - 212 ppm) and carboxylic acid group (190 – 203 ppm) are indicative for oxidation reactions. The autoxidation of aldehyde is summarized in Figure 75.

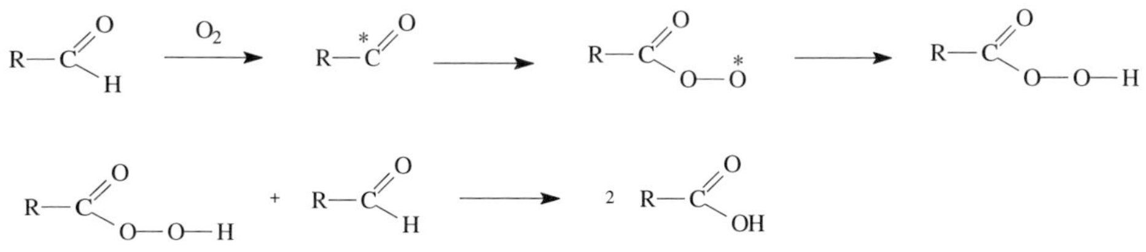

Figure 75: Autoxidation of aldehyde

The intensity for the methoxy signal is until the process step of the lignin fiber (LF) relatively. Considerable changes occur during the oxidation of lignin fiber (LF). For OLF the aliphatic (10 – 40 ppm) and the side-chain regions (70 – 90 ppm) completely disappear due to progressed oxidation. The intensity of the methoxy signal around 55 ppm decreases indicating that demethoxylation is one of the reactions during the oxidation (Figure 76).

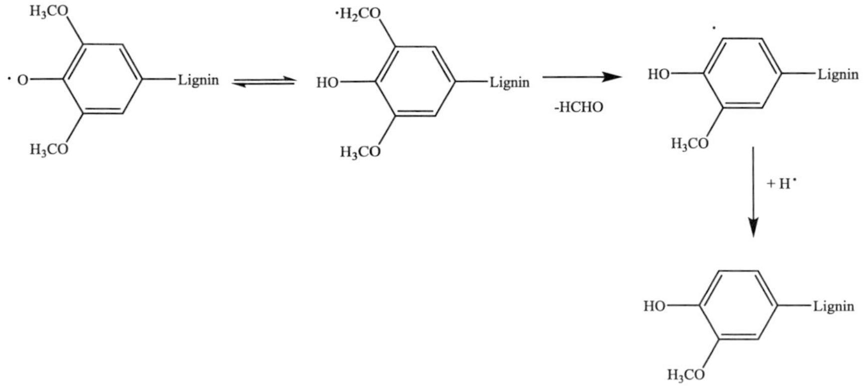

Figure 76: Demethoxylation

A very broad signal was detected for Aryl C-H and Aryl C-R appoximately 100 – 138 ppm this is indicative for non-specific changes in chemical environment of the aromatic carbons. This region is indicative for protonated and quartarnary carbon atoms.

Beste et al. have done molecular dynamic simulations of the reactions at the side-chains of lignin during conversion to carbon fiber [6-17]. The formation of phenantrenes was calculated in this simulation but never experimental proven. The ^{13}C chemical shifts measured in our experiments prove the formation of this chemical compound. The quartery carbon atoms of phenantrenes are known in the region between 130 and 132ppm and can be found in our experiments in this area. Furthermore significant changes were observed in the carbonyl region. An intensive signal was detected around 164 ppm and 190 – 200 ppm, representing anhydride and/or ester and aldehydes, respectively. The relative content of oxygenated structure, exclusively for aromatic structures of treated lignin samples, could be described using the integral of carboxyl and Aryl C-OR (225 – 139 ppm) in relation to the integral for Aryl C-H (135 – 100 ppm) [6-13]. For HW the integrals have a relation of 0.5, for P the integrals relation increase to 0.8 and for LF to 0.7. A very distinct improvement was calculated for OLF with 1.1. For a good signal-to-noise ratio the measuring time for OLF was with 30,000 scans relatively long. The origin of the low signal-to-noise ratio causes in the polyaromaticity within the sample [6-14]. It has been discussed that for higher temperatures and longer pyrolysis times polyaromatic hydrocarbons like phenanthrene, naphthalene and anthracene were formed [6-15]. Only one signal around 165 ppm was observed for the carbonized lignin fiber CLF (number of scans 70,000), associated with anhydride and ester. All other signals disappeared. An explanation for this phenomenon could be the complete cross-linkage of the lignin monomers. The formed polyaromatic structures are not measurable because the presence of paramagnetic species which results in complex relaxation behavior and shielding effects for the interested nuclei in the 13C CP/MAS experiment [6-16].

6.2.2 Fourier Transform Infrared Spectroscopy

The FT-IR spectroscopy is an effective method to describe the structural changes occurring during treatment. Figure 77 shows the FT-IR spectrum of Hardwood Lignin.

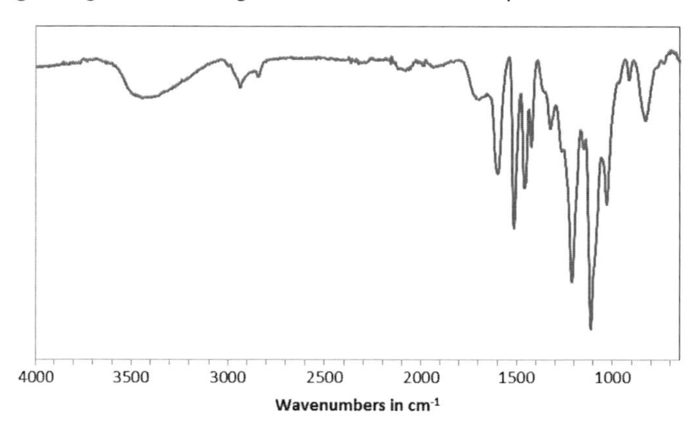

Figure 77: Figure 8: FT-IR spectra from 4000 – 650 cm-1 of Hardwood Lignin

The typically functional groups were found and the assignment was according the literature [6-6, 6-7]. Following bands were identified: O-H stretching at 3400-3460 cm^{-1}, aromatic C-H stretching 3009 cm^{-1} C-H stretching at 2937 and 2846 cm^{-1} in methoxy groups and methyl and methylene groups arising from side chains. Interesting region in the band occurs around 1705 cm^{-1} coming from C-O stretching in unconjugated carbonyl/carboxyl groups. The HW sample consists of a composition of syringyl and guaiacyl units. For both aromatic units the typical aromatic skeleton vibrations were found at 1600, 1514 and 1424 cm^{-1} and the aromatic ring vibration at 1462 cm^{-1}. The band at 1326 cm^{-1} and 1270 cm^{-1} are the ring breathing with C=O stretching and characteristic for S and G units, respectively. The band at 1113 cm^{-1} dominates and is associated with the C-H in plane deformation vibrations of the S units. The intensive band at 1214 cm^{-1} can be combined with the C-C, C-O and C=O stretching and the band at 1031 cm^{-1} with the C-H in plane deformation of guaiacyl units. The smaller band at 831 cm^{-1} is remarkable and in the range of aromatic C-H out-of-plane deformation vibrations for the S units.

Figure 78 shows the FT-IR spectra of the process line. A comparison between the HW, P and the fiber LF shows that the bands are rather similar and so the chemical structures are relatively equal. Due to the experimental method ATR the amount of samples used in IR could not be kept constant and so only the relative intensities of bands within a sample can be discussed. The O-H, C-H aromatic and aliphatic stretching bands (3400 – 2840 cm^{-1}) and the bands in the fingerprint region (1600 - 700 cm^{-1}) are identical. With growing processing level the C-O stretching band at 1705 cm^{-1} increased.

Generally, for the oxidized matt OLF a reduction of sharpness were observed for all bands. Very strong vibrations were detected for the C=O stretching band at 1705 - 1718 cm^{-1}, aromatic skeleton vibrations 1600 – 1420 cm^{-1}. In the region from 1400 - 1000 cm^{-1} a very broad signal without individual peaks were discernible. The out-of-plane aromatic deformations shifted to 770 – 750 cm^{-1}. This is indicative for changes in aromatic substitutions.

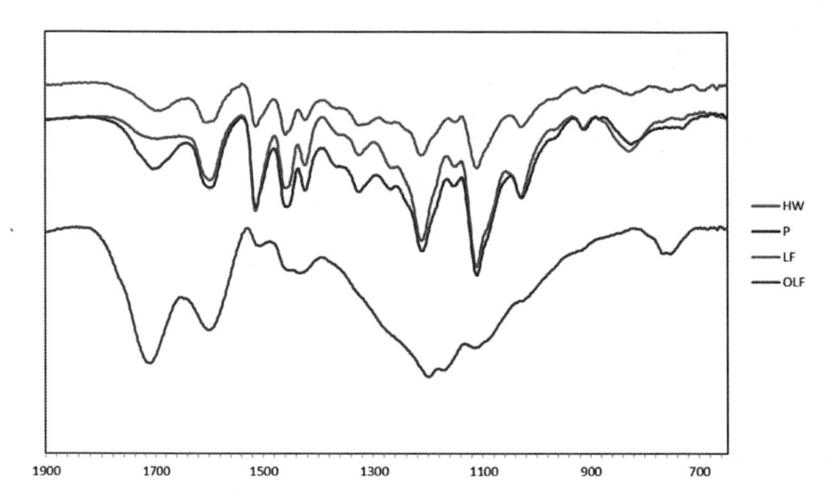

Figure 78: FT-IR spectra from 1900 – 650 cm^{-1} of the raw material the hardwood lignin HW , the pellets P, the fiber LF and the oxidized lignin matt OLF

6.2.3 Elementary Analysis

The used Hardwood Lignin has a carbon content of around 60%. For the pellets the carbon content was measured a little bit higher. The oxidized carbon fiber has, as expected, the highest oxygen content, which is an indication for successful crosslinking of the linomonomeres. The highest carbon content of 97% was measured in the carbonized fiber. That means the carbonization was successful performed. Table 18 summarizes the results of the elementary analysis.

Table 18: Results of the elemental analysis of HW, P, LF, OLF and CLF

Element	HW	P	LF	OLF	CLF
C	61.44	67.13	66.41	55.33	97.31
H	6.16	6.05	6.15	3.44	2.26
O	32.40	26.82	27.44	41.15	0.43

6.3 Major Reaction of Lignin during Conversion Process

The following section will show the major reaction of lignin during conversion process. On the basis of the results of the measurements presented in section three, chemical reactions for the pelletizing process, the fiber spinning process and the stabilization process are proposed.

6.3.1 Pelletizing of Lignin

The production of the pellets was done as described above. In the HSQC experiments following observations were made: On one hand the signals for the phenylcoumaran B2',6' as well as for S2',6' are not detectable anymore. On the other hand new signals for olefinic bonding are detected. The following reactions are postulate:

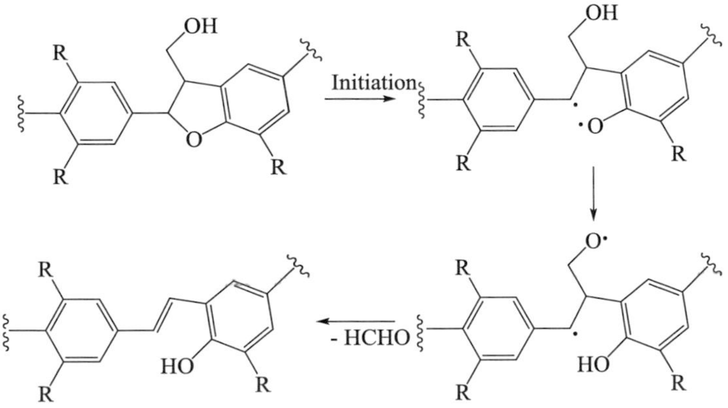

Initiation

- HCHO

Figure 79: Postulated reaction mechanism

Further reactions are the formation of carboxylic acids or ketone, according Figure 79. The FT-IR spectrum and the CP-MAS experiment support this postulated reaction. In the FT-IR spectrum the band at 1706cm^{-1} in the spectrum of the pellets is much stronger the in the spectrum of lignin. The CP-MAS shows nuclear magnetic resonance for the reaction of the aldehyde. The Elementary analysis shows also a higher carbon content of the pellets than the lignin powder.

6.3.2 Spinning of the Lignin Fiber

During the melt spinning process no major chemical reaction could be detected with FT-IR and CP-MAS. The detected spectrums are almost identical, which leads to the assumption, that no major chemical reaction is taking place during fiber spinning, since the fiber spinning is done only a little bit above glass transmission temperature.

6.3.3 Stabilization of the Lignin Fiber

Four chemical reactions are detectable during stabilization process:

- Formation of ketones
- Formation of carbon acids / Autoxidation of aldehydes
- Formation of cross-linkages

Formation of ketones

In the CP-MAS Spectrum of the oxidized fiber no aliphatic side-chains are detectable. The aliphatic side-chains were dismantled by oxidation reactions. An example for these reactions is the formation of ketones, which can detected at 190ppm in the ^{13}C CP-MAS.

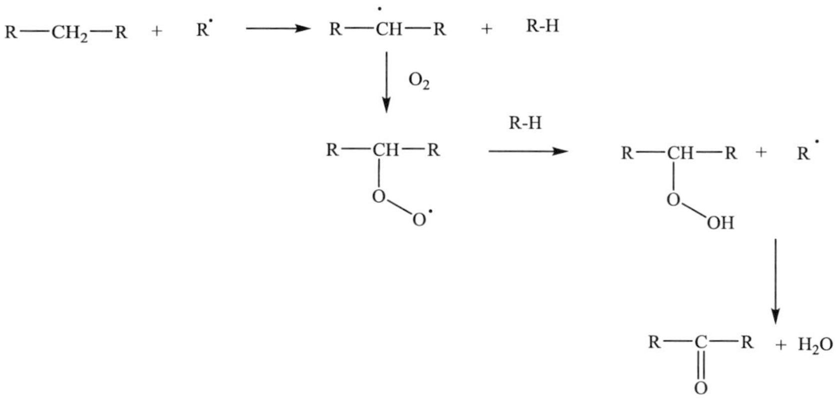

$$R{-}CH_2{-}R + R^{\bullet} \longrightarrow R{-}\dot{C}H{-}R + R\text{-}H$$

$$R{-}\dot{C}H{-}R \xrightarrow{O_2} R{-}CH(O{-}O^{\bullet}){-}R \xrightarrow{R\text{-}H} R{-}CH(O{-}OH){-}R + R^{\bullet}$$

$$R{-}CH(O{-}OH){-}R \longrightarrow R{-}C({=}O){-}R + H_2O$$

Figure 80: Postulated reaction mechanism – Formation of ketones

The proof of all reactions, presented in Figure 80, can also be found in the IR spectra. The rising of the signal at 1700cm^{-1} is characteristic for the C=O valence vibration, which is directly connected to the formation of ketones, carbon acids, aldehydes and anhydrides.

Formation of carbon acids / autoxidation of aldehydes

In the ^{13}C CP/MAS measurements a strong signal in the region between 190 and 200ppm is detectable. This signal is an indicator for the reaction of aldehydes, which creates carbon acids as described in Figure 79.

Formation of cross-linkages

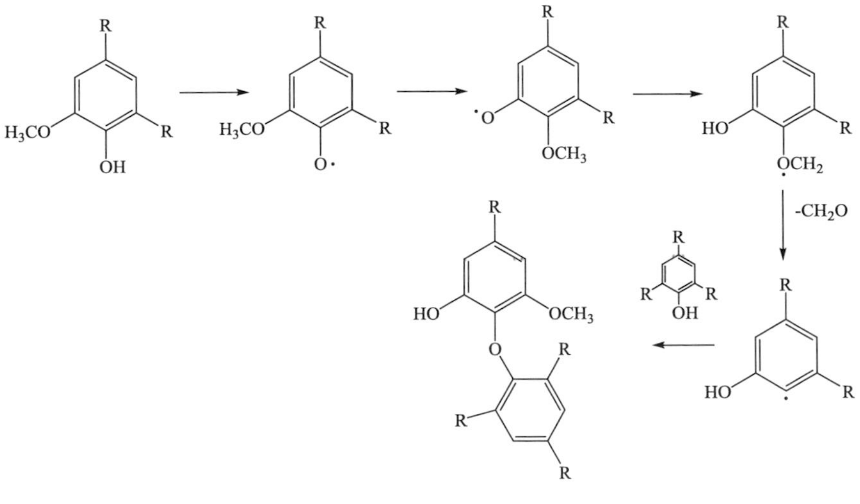

Figure 81: Formation of cross-linkages

In FT-IR spectrum the region from 1400 -1000 cm^{-1} a very broad signal without individual peaks was discernible. The explanation could be the increased cross-linked of the lignin macromolecules as described in Figure 81.

The results of the elementary analysis show a decrease of the hydrogen content and an increase of the oxygen content in the OLF. A further increase of the carbon content is detectable in the CLF.

6.3.4 Carbonization of the Lignin Fiber

The formation of the graphitic structure of the lignin based carbon fiber can be proven by the used analytical tools.

The detected signal at 165 ppm in the ^{13}C CP/MAS spectrum is a quarterly carbon atom and describes relicts at the aromatic rings. This relict contains oxygen, which are left over from reactions of anhydrides or esters (compare Figure 74).

The graphitic structure itself is not detectable with ^{13}C CP/MAS, but the high carbon content of 97% measured with the elementary analysis is a strong indication for the graphitic structure.

6.4 Conclusion

The NMR spectroscopy and the FT-IR spectroscopy are useful tools for analyzing the conversion process of lignin to carbon fiber.

The major reactions could be postulated by detecting of specific structural elements. During the pelletizing process the creation of double bonding carbon and carbonyl bonding could be established. The fiber spinning process is taking place without any detectable major reactions. In the stabilization process of the fiber the major reaction support cross linking in the fiber, which leads to higher mechanical properties. Also the creation of carbon acids, ketones and anhydrides is detectable in this step of the conversion process. In the carbonized fiber the graphitic structure can be postulated.

The knowledge of the reactions during the conversion process makes is now possible to optimize the production process of lignin based carbon fiber, which will lead to higher mechanical properties in the future.

6.5 References

[6-1] J.-L. Wen, S.-L Sun, B.-L XueR-C. Sun. Recent Advances in Characterization of Lignin Polymer by Solution Nuclear Magnetic Resonance (NMR) Methodology. Materials 2013; 6: 359-391.

[6-2] J. Ralph, L.L. Landucci. NMR of Lignins. In Lignin and Lignans: Advances in Chemistry; C. Heitner, D.R. Dimmel, J.A. Schmidt, Eds.; CRC Press: Boca Raton, FL, USA, 2010: 137-234.

[6-3] M. Foston, G.A. Nunnery, X. Meng, Q. Sun. NMR a critical Tool to Study the production of carbon fiber from lignin, Carbon. 2013; 52: 65-73.

[6-4] Argyropoulos D.S. Quantitative Phosphorus-31 NMR Analysis of Six Soluble Lignin. J.Wood Chem. Technol., 14:1; 1994: 65-82.

[6-5] Granata A., Argyropoulos D.S. 2-Chloro-4,4,5,5-teramethyl-1,2,3-dioxaphospholane a reagent for the accurate determination of uncondensed and condensed phenolic moieties in lignins, J. Agric. Food Chem. 1995; 46:1538-44.

[6-6] Boeriu, CG, Bravo, D, Gosselink, RJA, van Dam, JEG. Charakterisation of structure-dependent functional properties of lignin with infrared spectroscopy. Ind. Crops and Prod. 2004; 20: 205-218.

[6-7] O. Faix, O. Beinhoff, FTIR spectra of milled wood lignins and lignin polymer models with enhanced resolution obtained by deconvolution, J. Wood Chem. Technol. 1988; 8(4): 505-522.

[6-7] G.R. Hatfield, G.E. Maciel, O. Erbatur, G. Erbatur. Qualitative and Quantitative Analysis of Solid Lignin Samples by Carbon-13 Nuclear Magnetic Resonance Spectroscopy. 1987; 59: 172-179.

[6-8] G. Almendros, a.T. Martinzz, A.E. Gonzàlez, F.G. Gonzàlez-Vila, R. Fründ, H.-D. Lüdemann. CPMAS 13C NMR Study of Lignin Prparations from Wheat Straw Transformed by Five Lignocellulose-Degrading Fungi. J. Agric. Food Chem., 1992; 40: 1297-1302.

[6-9] E.A. Capanema, M.Y. Balakshin, J.F. Kadla. A Comprehensive Approach for

Quantitative Lignin Characterization by NMR Spectroscopy. J. Agric. Food Chem., 2004; 52:1850-1860.

[6-10] E.A. Capanema, M.Y. Balakshin, J.F. Kadla. Quantitative Characterization of a Hardwood Milled Wood Lignin by NMR Spectroscopy. J. Agric. Food Chem., 2005; 53:9639-9649.

[6-11] J.L. Braun, K.M. Holtman, J.F. Kadla. Lignin-based Carbon Fibers: Oxidative Thermostabilization of Kraft Lignin. Carbon. 2005; 43:385-394.

[6-12] Y. li, D. Cui, Y. tong, L. Xu. Study on structure and thermal stability properties of lignin during thermostabilization and carbonization. Int. J. bio. Macromol., 2013; 62: 663 – 669.

[6-13] B. D. Diehl, N. R. Brown, C.W. Frantz, M.L. Lumadue, F. Cannon. Effects of pyrolysis temperature on the chemical composition of refined softwood and hardwood lignin. Carbon, 2013; 60:531-537.

[6-14] M. Asmadi, H. Kawamoto, S. Saka. Pyrolysis reactions of Japanese beech woods in a closed ampoule reactor. J. Wood Sci., 2010; 56: 319-330.

[6-15] J.A. Fran z, R. Garcia, J.C. Linehan, G.D. Love, C.E. Snape. Single-pulse excitation 13C NMR measurements on the Argonne premium coal samples. Energy Fuels, 1992; 6:598-602.

[6-16] A. Beste, ReaxFF Study of the Oxidation of Lignin Model compounds for the Most Common Linkages in Softwood in View of Carbon Fiber Production. J. Phys. Chem. 2014; 118: 803 – 814.

[6-17] Y. Li, D. Cui, Y. Tong, L. Xu. Study on structure and thermal properties of lignin during thermo stabilization and carbonization. Int. J. Bio. Macromolecules, 2013; 62: 663-669

7 Properties and Chemical Structure

The lignin based carbon fibers, which were produced with the procedure presented in chapter 5, are analyzed in this chapter. Using several types of spectroscopy such as Raman and X-ray photoelectron, the change of the chemical structure of the fiber during the conversion process was detected. Properties like tensile strength, the elongation, and the Young's modulus were measured in the single fiber tensile test.
The potential of Hardwood Lignin as a precursor for carbon fiber is analyzed by combining the results of the properties and chemical structure.

7.1 Properties of Lignin Based Carbon Fiber

To obtain the mechanical properties single fiber tensile tests were performed. The density and the porosity were analyzed using helium gas pycnometry and a scanning electron microscope.
The mechanical properties of the lignin based carbon fibers are highly influenced by the microstructure of the fiber and the defects in the fiber. The defects are responsible for local excessive tension in the lignin based carbon fiber, which lowers the mechanical performance of the fiber. The used method for obtaining images of the fiber microstructure was scanning electron microscopy.

7.1.1 Single Fiber Tensile Test

The tensile test is a destroying material testing procedure and is used in this context for the detection of the mechanical properties of the lignin based carbon fiber. The single fiber tensile test offers the opportunity to measure the mechanical properties of a fiber without any influence of effects of a composite such as fiber-matrix bonding, or fiber damage from handling during production of the textile or the composite.

Theoretical Background

The detected mechanical properties which were measured with the single fiber test are: the tensile strength, the elongation, and the Young's modulus [7-1]. The detection of the strength [σ] is done with the following equation 7-1. The maximum force [F] divided by the area cross-section [A] equals the strength [σ] [7-2].

$$\sigma = \frac{F}{A} \tag{7-1}$$

To calculate the elongation [ε] (equation 7-2), the change of the length of the fiber [Δl] is divided by the original length of the fiber [l] [7-2].

$$\varepsilon = \frac{\Delta l}{l} \tag{7-2}$$

The Young‘s modulus [E] follows the equation 7-3, where the elastic relation strength and elongation is described [7-2].

$$E = \frac{\sigma}{\varepsilon} \tag{7-3}$$

Experimental

The single fiber tensile test measures the mechanical properties of the lignin based carbon fibers. The norms DIN EN ISO 291 and DIN EN ISO 11566 were used to make the results comparable to mechanical properties of other carbon fibers.

Before the single fiber test can be performed, the fibers were conditioned following the description found in DIN EN ISO 291 [7-3]. The fibers were stored for 24 hours at a relative humidity of 50 % and a temperature of 23 °C. The climate in the testing room was similar to these conditions.

The fiber testing was done with testing equipment Favimat+ from the company Textechno (Germany). The length of the filament and the testing speed are given in DIN EN ISO 11566 [7-4] and summarized in Table 19. The yarn count T was determined with the vibration method following DIN EN ISO 1973. Figure 82 shows how the vibration method is performed. The filament is clamped with initial load and sound waves vibrate the filament. The detected frequency is proportional to the yarn count.

Table 19: Parameter of the single fiber tensile testing [7-5]

	Parameter	Unit
Density	1.8	g/cm³

Length of the filament	25	mm
Testing speed	5	mm/min

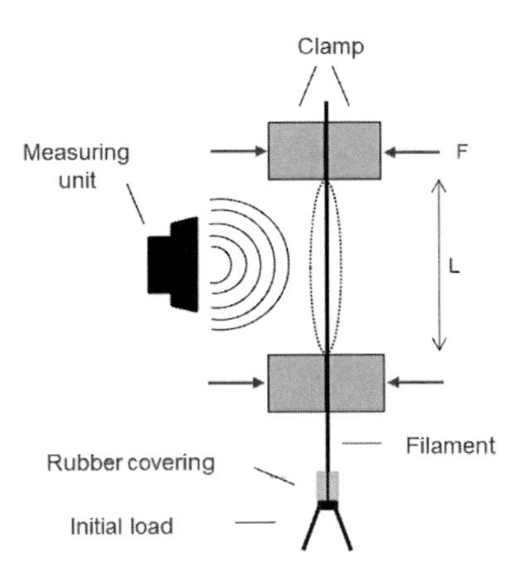

Figure 82: Schematically illustration of the single fiber tensile testing equipment [7-5]

Results and Discussion

The analysis of the measurement results was done as described in DIN EN ISO 11566. For determining the strength, the cross-section [A] was calculated with the equation 7-4 given in DIN EN ISO 1973.

$$A = \frac{T}{\rho} \quad (7\text{-}4)$$

The yarn count T was determined with the vibration method following DIN EN ISO 1973 [7-4]. Figure 82 shows how the vibration method is performed. The filament with a length [l] is clamped with initial load and sound waves vibrate the filament. The detected frequency is proportional to the yarn count. With the detected frequency [f] the yarn count of the filament is calculated with equation 7-5 [7-6].

$$T = \frac{F}{4 * f^2 * l^2} \quad (7\text{-}5)$$

The results from the measurements are summarized below. Table 20 shows the peak stress, the modulus and strain of the stabilized lignin fiber and Table 21 presents the same parameters of the lignin based carbon fiber.

Table 20: Single fiber tensile data of stabilized lignin fiber

Sample	Diameter [µm]	Peak stress [MPa]	Modulus [Gpa]	Strain at peak stress [%]
1	21.29	31.03	1.38	1.85
2	19.76	21.37	0.69	1.71
3	11.44	17.24	0.69	4.37
4	13.87	18.62	2.07	0.91
5	20.88	23.44	2.07	0.97
6	20.64	23.44	2.07	1.51
7	17.59	41.37	2.76	0.97
8	11.40	42.06	4.14	1.17
9	11.49	57.23	3.45	1.10
10	14.54	23.44	3.45	1.76
Mean	**16.29**	**29.92**	**2.28**	**1.60**
Std. Dev.	**3.97**	**12.30**	**1.11**	**0.98**

A comparison of the tensile data of the two fibers shows how the conversion process effects the mechanical properties. The diameter reduces from around 16 micron to approximately 12.5 microns. The peak stress is increased by fivefold, and a modulus that is ten times greater. This data also shows that the conversion process was done successfully.

Table 21: Single fiber tensile data of carbonized lignin fiber

Sample	Diameter [µm]	Peak stress [MPa]	Modulus [GPa]	Strain at peak stress [%]

1	14.98	111.70	17.24	0.63
2	12.11	243.38	22.06	0.93
3	12.38	109.63	17.24	0.59
4	15.49	103.42	21.37	0.37
5	11.79	90.32	19.99	0.45
6	12.04	119.97	22.75	0.52
7	12.73	125.48	14.48	0.40
8	11.25	277.17	24,13	1.15
9	12.61	124.11	22,75	0.52
10	10.89	204.77	20.68	0.92
Mean	**12.63**	**151.00**	**20.27**	**0.65**
Std. Dev.	**1.41**	**62.36**	**2.90**	**0.25**

The mechanical properties are connected to a high porosity (Chapter 7.1.2 and Chapter 7.1.3) and to a high crystallinity of the lignin based carbon fiber (Chapter 7.2).

7.1.2 Density of Lignin Based Carbon Fiber

“Pycnometry” is derived from the Greek word puknos, which has long been identified with volume measurements. The helium pycnometer is specifically designed to measure the true volume of solid materials by employing Archimedes’ principle of fluid displacement and gas expansion (Boyle's Law) [7-7].

Theoretical Background

Ideally, a gas is used as the displacing fluid since it penetrates the finest pores ensuring maximum accuracy. For this reason, helium is recommended. Since helium has the smallest atomic dimension, it enables the entry into crevices and pores approaching 0.2 nm. Its behavior as an ideal gas is also desirable. Other gases such as nitrogen can be used, often with no measurable difference [7-8, 7-9, 7-10].

Experimental

All measurements were made with the Ultrapyc 1200 from Quantachrome Instruments. All density measurements were done with the maximum number of 99 runs. The run data can be found in Appendix 12.1.

The following analysis parameters were used:

- Temperature: 23.9 °C
- Type of gas: helium
- Volume cell: 9.8584 cm^3
- Volume added: 8.4859 cm^3

Results and Discussion

Table 22 shows the results of the measurements of the density of the lignin fiber, stabilized lignin fiber, and carbonized lignin fiber (compare production process of lignin based carbon fiber in Chapter 5)

Table 22: Analysis results of the density of lignin based carbon fiber

	Lignin Fiber	Stabilized Lignin Fiber	Lignin based Carbon Fiber
Sample Weight [g]	0.4861	0.5970	0.6253
Average Volume [cm³]	0.3762	0.3931	0.32.80
Volume Std. Dev. [cm³]	0.0007	0.0019	0.0032
Average Density [g/cm³]	**1.2921**	**1.5188**	**1.9064**
Density Std. Dev. [g/cm³]	0.0026	0.0073	0.0183

The results of the density measurements shows an increase in the density due to the conversion process. This can be proven with Raman and XPS measurements (Chapter 7.2.1 and 7.2.4) which show the rising of carbon content in the fiber. The results of the tensile tests correlate to these results since the strength of the fiber is rising during conversion process.

7.1.3 Scanning Electron Microscopy

The scanning electron microscope (SEM) was used to show the surface and friction area of the lignin based carbon fiber. The SEM can be used to show the porosity of the fiber and some representative defects.

Theoretical Background

Figure 83 shows how a SEM principally works [7-11, 7-12].

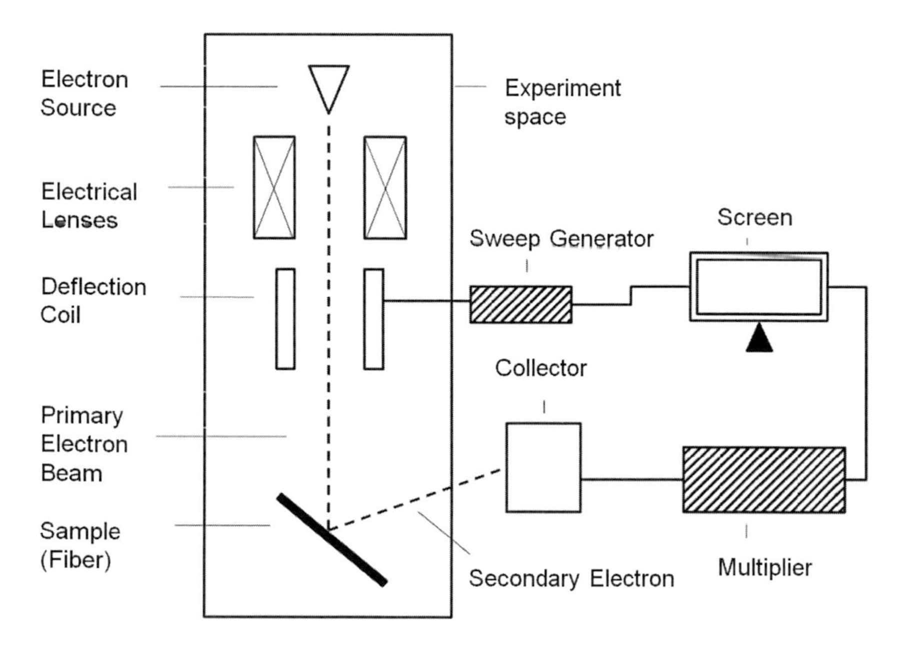

Figure 83: Principal setting of a SEM [7-11, 7-12]

The primary electron beam is made by heating a wolfram cathode and speeding up the electrons through an anode. Before scanning the surface of the sample, the electron beam is focused using the electrical lenses. The defection coil is synchronized with the screen to control the scanning process [7-12]. During scanning of the surface, two kinds of new electrons are produced: reflected electrons (Energy > 50eV) and secondary electrons (Energy < 50eV). The secondary electrons are collected from a detector and creating the picture of the surface of the fiber [7-13]. A resolution in the nanometer scale with an enormous depth of focus is possible using XPS [7-14].

Experimental

To obtain the SEM measurements the high resolution SEM Zeiss Ultra 55 was used. It contains a standard Eberhart-Thornley secondary electron (SE) detector, an In-Lens SE detector, and a EsB (energy and angle selective backscattered electrons) detector. The microscope can work with an image enlargement range of 12,000 to 900,000 times with the SE detector, and an image enlargement range of 100,000 to 900,000 times with the EsB detector. The nominal resolution is about 1 nm, and even with low increase in speed, an electric tension of only 1 kV a resolution of 1.7 nm is possible. With the (In-Lens) EsB detector, material contrast can be shown. The signal from Se and BsE detector can be shown on two screens or overplayed and mixed, so that high resolution and high contrast images are available.

To perform the SEM imaging of the fibers, the fibers needed to be prepared on a sample holder for measurement. For the preparation of the fibers, concepts using a carbon pad and focusing the fibers as a bale were tested. Clamping the fibers using carbon pads offers the best way to create high resolution images. After that the sample was coated with gold using the physical vapor deposition (PVD) process. To obtain high resolution images without any influence of vibration, the fiber length was chosen as short as possible above the clamping point.

To make the high resolution images, an overview image was taken first and then representative fibers were chosen for making the high resolution images itself. Also the combination of different detectors, the optimal distance of the sample, and different increase in speed electric tension was evaluated to get the best contrast in the images. The best parameters were at a distance of 5mm, an increase in speed electric tension of 5 kV and the use of the In-Lens SE detector. Overview images had a typical resolution of 5000 times. Detail images were captured with a resolution of 15000 to 20000 times.

Results and Discussion

Lignin Fiber

The structure of the lignin fiber is partly irregular. The fiber cross section is round with different diameters in a range between 10 and 25 µm. The fiber surface is smooth with some pores. The fiber cross section shows round voids which are mostly located at the fiber edge (Figure 84).

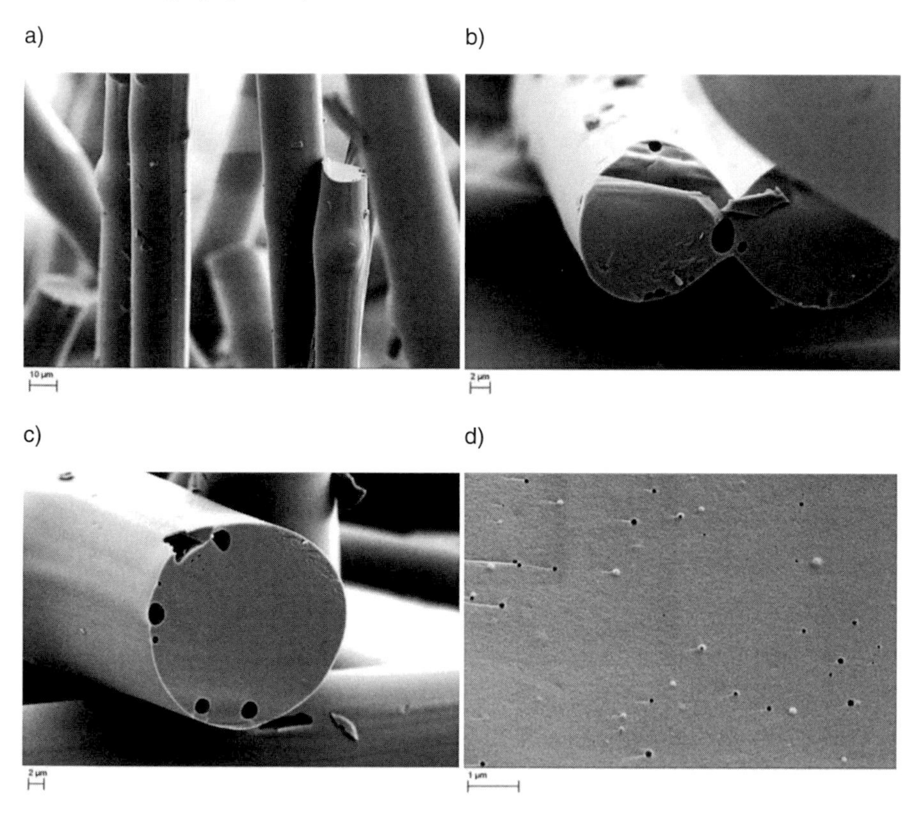

Figure 84: SEM images of lignin fiber: a) overview image, b) fused fibers, c) voids at the fiber edge, d) pores at the surface

The reason for the round defect in the lignin fiber are vaporously components in the lignin, which vaporize during melt spinning and create air bubbles in the fiber (compare Chapter 5). This defects cause lower mechanical properties.

Stabilized Lignin Fiber

After the stabilization process (compare chapter 5.4), the fiber diameter is reduced. The voids created during the melt spinning process don't change in size causing the ratio between voids and cross-sectional area to decrease. A higher inhomogeneity concerning the fiber diameter is detectable (Figure 85). It is noticed that the fiber surface is smooth and the pores at the surface are greatly reduced.

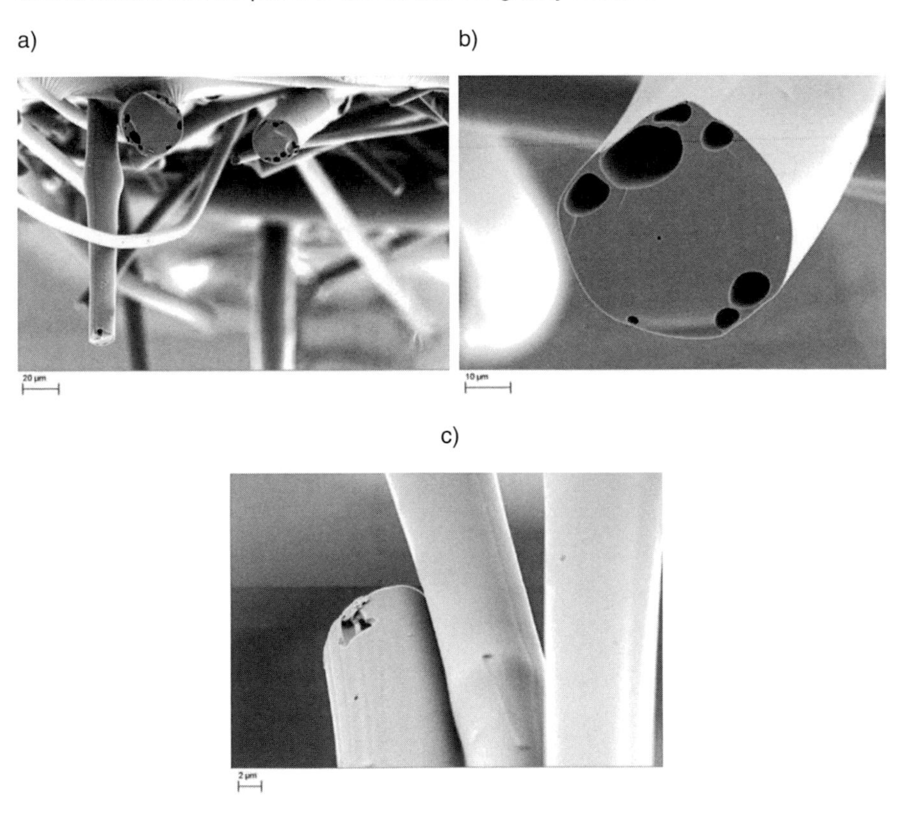

Figure 85: SEM images of stabilized lignin fiber: a) overview image, b) defects at the edge, c) defects at the surface

Lignin based Carbon Fiber

The lignin based carbon fiber shows the same defects like the lignin fiber and the stabilized lignin fiber (Figure 86). After the carbonization process (compare chapter 5.4), the fiber diameter is reduced. The voids created during the melt spinning process don't change in size causing the ratio between voids and cross-sectional area to decrease.

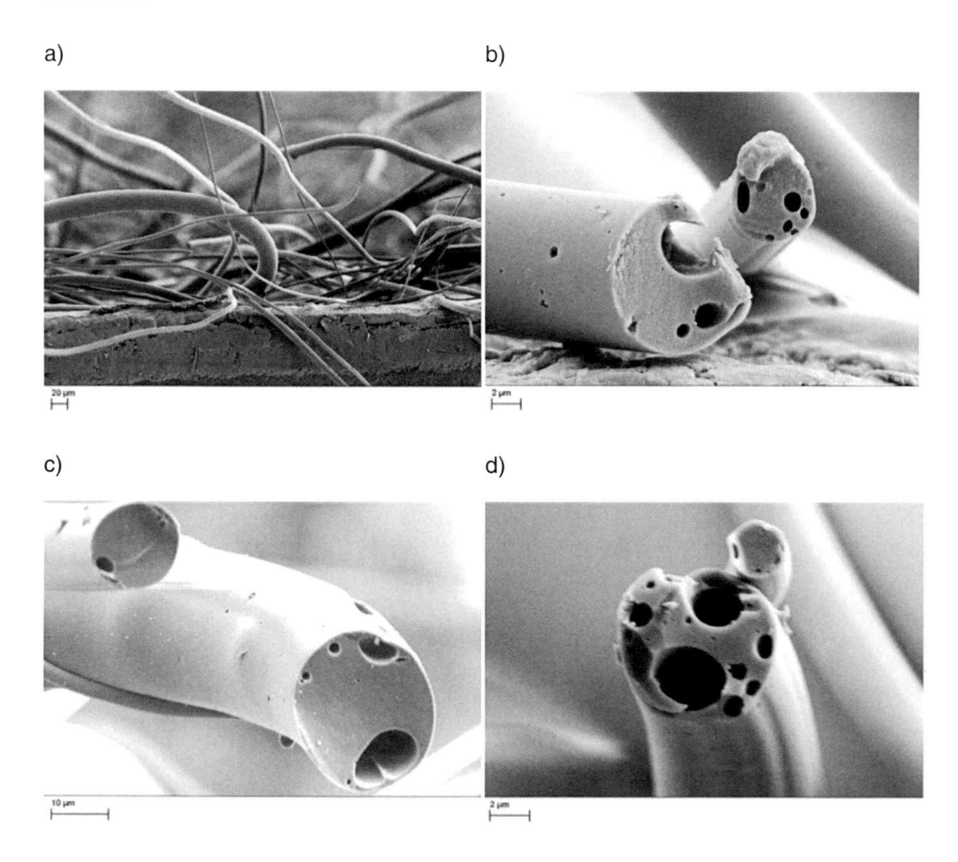

Figure 86: SEM images of lignin based carbon fiber: a) overview image, b) defects at the surface, c) defects at the cross section, d) defects and inhomogeneity

The heat treatment of the lignin fiber leads to a reduction of the fiber diameter, since all other elements except carbon is removed from the fiber to create a lignin based carbon fiber (compare Chapter 7.2.1 and Chapter 7.2.4). The mechanical properties improve during conversion process and the fiber becomes less elastic and more brittle (for the chemical reasons compare Chapter 7.2.1).

Comparison of the Area of Fracture

Using samples from the single fiber testing, the area of fracture for both the lignin fiber as well as the lignin based carbon fiber was analyzed with SEM. Figure 87 shows the area of fracture of both fiber types.

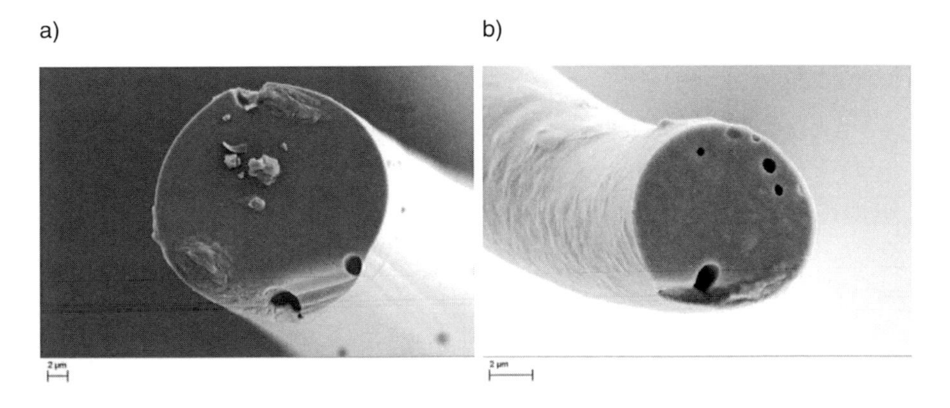

Figure 87: Comparison of the area of fracture: a) lignin fiber, b) lignin based carbon fiber

The lignin fiber exhibits a ductile fracture. This is an indication for a more elastic fraction behavior. The lignin based carbon fiber shows a brittle fraction behavior since the fraction area is very even without any deformation. Chapter 7.2 will help to explain this phenomenon with the help of the chemical structure of the fibers.

7.2 Chemical Structure of Lignin Based Carbon Fiber

After detecting the mechanical properties and analyzing the macroscopic structure with the help of SEM, the chemical structure of the lignin based carbon fiber will explain how the properties of the fiber increases during the conversion process.

7.2.1 Raman Spectroscopy

The Raman spectroscopy makes it possible to detect the bonding of carbon in the fibers. Using Raman spectroscopy provides the measurement of the quantity of sp^2 and sp^3 hybridization of the carbon atoms in the fiber. The measurements obtained can show the chemical transformation of the lignin fiber to the lignin based carbon fiber. This data can also show how the relationship between graphite (sp^2) and diamond (sp^3) bonded carbon correlates to the fibers mechanical performance.

Theoretical Background

The Raman Effect is the fundamental physical phenomenon, which is used to perform the Raman spectroscopy. When a sample is radiated with photons from a high intensity laser, the molecules begin to vibrate [7-15, 7-16, 7-17]. Only a small part of the radiation is interacting with the molecules, causing the Raman Effect. Figure 88 shows the energy levels of the different light scattering possibilities. The dashed line shows a virtual energy level [7-18].

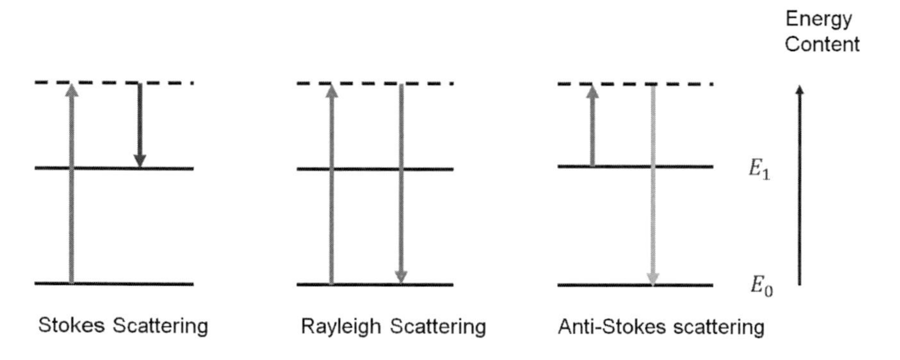

Figure 88: Raman effect: possible energy levels

In the case of Stokes Scattering and Anti-Stokes Scattering the collision between the photon and the molecule is inelastic. In the case of Stokes Scattering, the molecule gains energy, while the photon loses energy. In the case of Anti-Stoke Scattering the molecule has already higher energy content and the energy is transferred to the photon [7-16]. The Rayleigh Scattering describes the elastic collision of a photon with a molecule, where the energy content is unchanged.

The energy content in the Raman spectra is diagramed with the help of the wavenumber [k] (Figure 88). The equation 7-6 shows how the wavenumber can be calculated by using the speed of light [c] (c =2,99792458*10^8 m/s) and the frequency [f] of the photon source [7-19].

$$k = \frac{f}{100*c} \tag{7-6}$$

For illustration purposes, the sp^2 and sp^3 hybridization is shown in Figure 88. The G-band (sp^2) is described as a ring vibration of the C6-ring in the carbon material [7-20]. The D-band (sp^3) is characteristically for a central vibration of a diamond structure. The direction of the sp^2-orbitals is symmetrically triangular and the sp^3-orbitals are tetrahedral.

a) sp^2 Hybridization

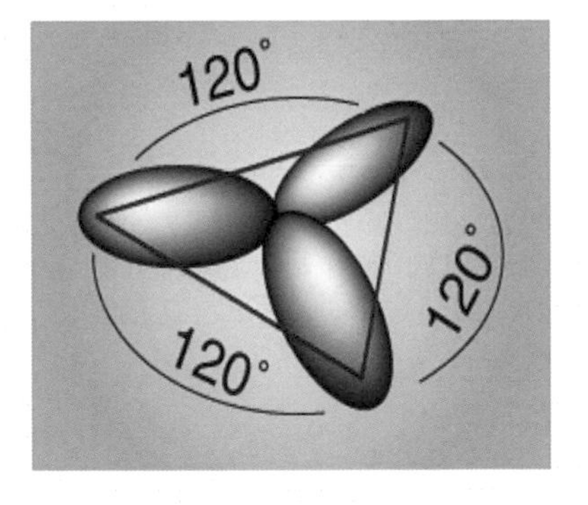

b) sp^3 Hybridization

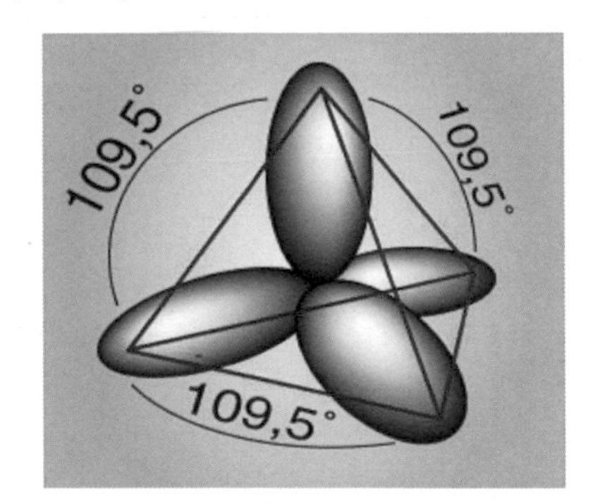

Figure 89: Hybridization of Carbon [7-20]

The angles between the sp^2-orbitals are 120 ° and the angles between the sp^3-orbitals are 109.5 °. Hybrid orbitals illustrate wave functions, which makes it possible to have a pictorial idea of the spatial shape of probability to meet an electron there [7-21]. That means three of the four electrons in the sp^2 hybridization are bonded in atomic bonding and one electron is not used for any bonding. This electron can be found between the graphite layers [7-22]. In the case of sp^3 hybridization all four electrons are part of atomic bonding, which means the carbon atom becomes the center of a triangular pyramid.

Experimental

To perform the Raman spectra measurments the Nicolet Almega XR from the company Therma Nicolet was used. It is a fully automatic software driven Raman spectrometer, with two lasers as the photon sources. The double frequency Neodymium Doped Yttrium Orthvanadate Diode Pumped Solid State Laser (532 nm, 25mW) and the external stabilized diode laser (780 nm, 25nW) offers the possibility to measure in two different spectral areas (7cm^{-1} and 2cm^{-1}). The 532 nm laser can work in low resolution mode with wave numbers 6616 to 4443 cm^{-1}. The high resolution mode of the same laser uses wave numbers in the range of 4090 to 710 cm^{-1}. The external stabilized diode laser (780nm laser) works in areas between 3098 to 3998 cm^{-1} for low resolution images, and uses wave numbers in the range of 4001 to 3098 cm^{-1} for high resolution measurements.There are 4 different lenses available to focus the laser beam (10x, 50x, 50x long distance and 100x), which makes it possible to detect areas of the sample in the range of 4 and 1 µm.

Before performing measurement the samples had to be prepared and a calibration of the lasers was necessary. Therefore the fibers were cut mechanically in 5 mm long specimens and fixed on a sample holder parallel to each other. The measurement itself was done in the chamber of the Raman spectroscope. This chamber is fully isolated against electromagnetic radiation, to prevent any influence on the spectra.

Several tests were made to verify, which laser system was more suited for this application. It is necessary to find the best combination of wavelength and energy of the laser. The wavelength of the laser can cause florescence of the sample, and the energy of the laser can convert the sample. It turned out that the 532 nm laser indicated too much florescence in the sample, which made it difficult to create clear spectra. For this reason all measurements were done with the 780 nm laser as the photon source. To prevent any damage of the sample by the laser, the energy and measurement time of the 780 nm laser needed to be tested. The high energy of the laser and long measurement time leads to conversion of the fiber. To ensure, that no conversion is taking place, the measurement times were chosen as short as possible and the laser energy was also used as low as possible. This testing creates a set of optimal parameters, which minimized signal to-noise ratio. The combination of 30 % laser power, use of the ten times lens, and an integration time of 60 s supplies the best spectra of the fibers. For each sample three spectra at three different areas of the sample were made. Every spectrum contains 20 single spectra, which were combined from the software to one spectrum.

Results and Discussion

The Raman spectra makes it possible to show sp^2 and sp^3 hybridization of the carbon atoms in the investigated fiber. Using Raman Spectroscopy the conversion of the fiber during carbon fiber production can be illustrated. The relation between the graphitic (sp^2) and diamond structure (sp^3) of the carbon is an indicator for the mechanical properties of the fiber. A high content of graphitic and diamond bonded carbon results in high mechanical properties.

The Raman spectra of the lignin fiber, the stabilized lignin fiber and the lignin based carbon fiber are shown in Figure 90. In the spectrum of the lignin fiber (Figure 90a) no bands are detectable, which means the lignin fiber shows a typical spectrum for a polymer. The Raman Spectra of the stabilized lignin fiber (Figure 90b) clearly shows the G-band at 1350 cm^{-1}. The formation of the G-band is an indicator for the successful performance of the stabilization process by oxidation reactions. The lignin based carbon fiber (Figure 90c) has a D and G-band. Both bands are clearly visible and show that the carbonization process operates successful.

a) Lignin Fiber

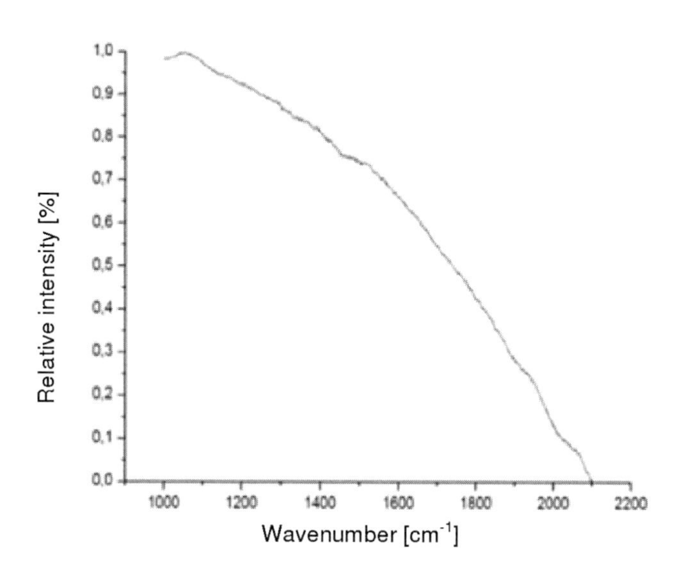

b) Stabilized Lignin Fiber

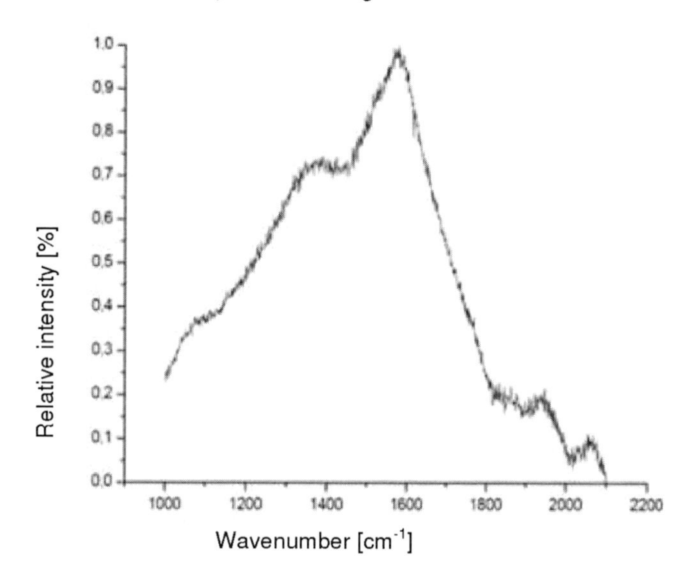

c) Lignin Based Carbon Fiber

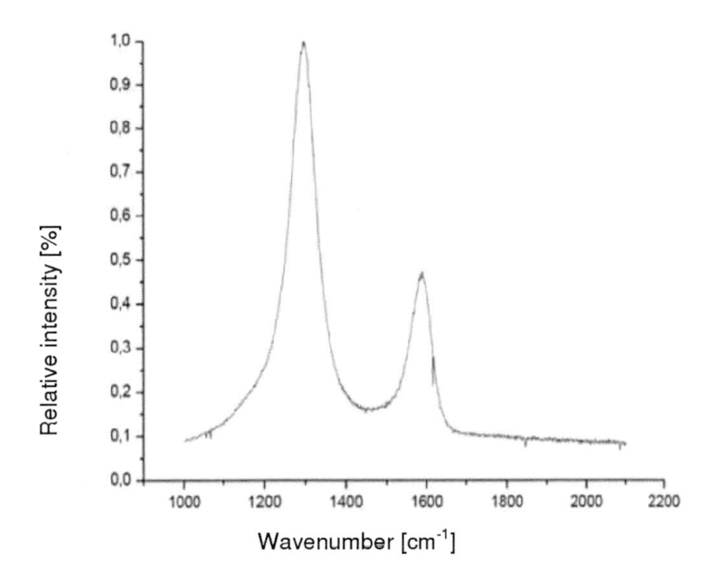

Figure 90: Raman Spectra: a) Lignin Fiber, b) Stabilized Fiber, c) Lignin Carbon Fiber

Characteristic for carbon atoms in the carbon fiber is the G-band, which illustrates the sp^2 hybridization, and the D-band, which illustrates the sp^3 hybridization. Both bands stand for the carbon modification of graphite (sp^2) and diamond (sp^3). The sp^2 hybridized G-band was typically found at wavenumber of $1350cm^{-1}$ and the sp^3 hybridized G-band at 1550 cm^{-1} [7-19].

A high content of sp^3 hybridization of the carbon bonding (diamond structure) causes a high stiffness and a high tensile strength. In the diamond structure of carbon every carbon atom is bonded to four other carbon atoms. All four outer electrons of a carbon atom are involved in atomic bonds to other carbon atoms, and all atoms have the same distance to each other. Four carbon atoms are bonded this way to the center carbon atom and create a triangular pyramid, which represent the smallest part of the highly regular diamond structure. The very stable atomic bonds in this crystalline structure are the reason for the high stiffness and the high tensile strength of carbon fibers [7-23, 7-24, 7-25].

The anisotropic properties of a carbon fiber can be explained by the sp^2 hybridization of the carbon atoms in the carbon fiber (graphite structure). In the graphite structure every carbon atom is bonded to three other carbon atoms. The resulting structure is a hexagon build from six carbon atoms. These hexagons create layers. The fourth outer electron of the carbon atom in sp^2 hybridization is not participating in the atomic bonds and can move between the layers, which explain the low bonding properties between these layers and the high stiffness and the high tensile strength in the layer [7-23, 7-24, 7-25].

A comparison of the Raman spectra of the lignin based carbon fiber and a conventional carbon fiber can be found in chapter 8.1.

7.2.2 X-ray Photoelectron Spectroscopy

X-ray photoelectron spectroscopy (XPS) is an important method for analyzing the chemical elements, chemical bonding, and electrochemical structure of a carbon fiber. XPS spectra are obtained by irradiating the fiber with a beam of X-rays while simultaneously measuring the kinetic energy and number of electrons that escape from the fiber being analyzed. This way the conversion of lignin to carbon fiber can be illustrated by the change of the elemental composition and the change of chemical bonding [7-26].

Theoretical Background

Electrons in a chemical bonding have a defined binding energy E_B, which defines the strength in a chemical bond. The binding energy is correlated to the quantum number n and the atomic number Z:

$$E_B = -13{,}6\text{eV} * \frac{Z^2}{n^2} \tag{7-7}$$

The binding energy between two atoms is unique. The kind of atoms and the kind of chemical bonding define the binding energy. Analyzing the binding energy is the basis of the chemical analysis, with the help of XPS [7-27].

Because the energy of an X-ray with particular wavelength is known, and because the emitted electron's kinetic energy is measured, the electron binding energy of each of the emitted electrons can be determined by using the photoelectrical effect and the equation presented in Figure 91. The photoelectron spectrometer detects these emitted electrons [7-28, 7-29].

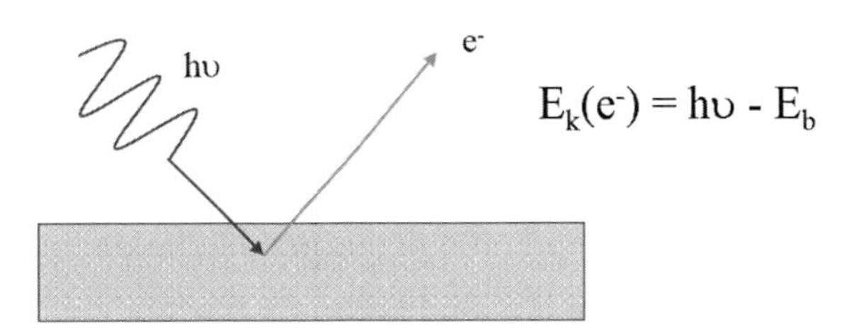

Figure 91: Schematically illustration of the photoelectrical effect [7-30]

A typical XPS spectrum is a plot of the number of electrons detected versus the binding energy of the electrons detected. Each element in a carbon fiber produces a characteristic set of XPS peaks at characteristic binding energy values. The number of detected electrons in each of the characteristic peaks is directly related to the amount of element within the XPS sample volume [7-31, 7-32].

Experimental

Using XPS (X-ray photoelectron spectroscopy) the elementary composition of the surface of the lignin based carbon fiber was analyzed. For the measurements a PHI 5600 LS spectrometer was used with non-monochromatic Mg-Kα-radiation (300 W, 13 kV, 24 mA). The used XPS had a depth of information less than 10 nm. The detection border was around 0.5 atom %. The pressure in the measurement chamber was below 10^{-8} torr (Ultra-High Vacuum).

Using the described XPS, overview and detailed spectra were performed. Overview-spectrums had a period time of 20 minutes and were measured in a range of

0 – 1100 eV. The passing energy of 190 eV and a step size of 0.4eV/step were used for the overview-spectrums. Detailed spectrums for the elements carbon and oxygen had a period time 20 min per element. They were made with pass energy of 30eV and a step size of 0.125eV/step. The spectrums were evaluated with the software CasaXPS. The software used a component fitting with a Gauss-Lorenz form and a Gauss-Lorenz-relation of 70/30. Since the fibers conduct electricity, they can get charged be electrons which did not leave the surface. The charging of the samples was corrected by fixing the Carbon-Carbon-signal (C-C) at 285 eV.

Results and Discussion

Overview spectrums and high resolution spectrums were two kind of measurements done for the lignin fiber, the stabilized lignin fiber and the lignin carbon fiber. The overview-spectrum tells the percentage of Carbon, oxygen and nitrogen in the fibers (Table 18). Based on the overview spectrums, the relation O/C was calculated (Table 19). The use of the detailed spectrum tells the percentage of carbon bonds to other atoms. The detailed spectrum also presents the number of aliphatic carbon-carbon-bonds. A high percentage of carbon-carbon bonds in the fiber are an indicator for high mechanical properties.

The following table presents all detected elements.

Table 23: Detected elements by XPS

Atom %	C	O	N
Lignin Fiber	80.5	19.5	0.0
Stabilized Lignin Fiber	72.1	27.9	0.0
Lignin based Carbon Fiber	94.2	5.8	0.0

The next table shows the relation between oxygen and carbon. The absolute error is smaller than ± 0.03, which means the determined high resolution spectrum correlates to these measurements.

Table 24: Detection of the O/C Relation by XPS

	Lignin Fiber	Stabilized Lignin Fiber	Lignin based Carbon Fiber
O/C	0.24	0.39	0.06

Detailed scans were made for the atoms oxygen and carbon. The carbon signal was fitted with five components (C-C, C-H, C=C / C-O, C-C-O / C=O, O-C-O / O-C=C, shake up*) and the oxygen signal was fitted with carbon. The resulting bonding-energy (E_B) and the bonding ratios (%) are shown in Table 25. The fitting interval was chosen with ≤0.5 eV.

Table 25: Detailed XPS scans of oxygen and carbon

Region	Class	E_B /eV	Lignin Fiber [%]	Stabilized Lignin Fiber [%]	Lignin based Carbon Fiber [%]
C1s	C-C	285.0	53.4	39.5	66.0
	C-H	285.0	53.4	39.5	66.0
	C=C	285.0	53.4	39.5	66.0
	C-O	286.5	24.0	21.9	14.3
	C-O-C	286.5	24.0	21.9	14.3
	C=O	288.0	0.5	0.7	4.1
	O-C-O	288.0	0.5	0.7	4.1
	O-C=O	289.0	1.7	8.8	7.0
	Shake up*	291.5	0.5	1.0	2.4
O1s	O-C	533.7	19.9	28.1	6.2
	C-O-C	533.7	19.9	28.1	6.2

O1s	C=O	533.7	19.9	28.1	6.2
	O-C-O	533.7	19.9	28.1	6.2
	O-C=O	533.7	19.9	28.1	6.2

* "Shake up" Carbon in aromatic compound (crossing π–π*)

Further evidence of the oxidation and condensation reactions taking place during the thermal treatment of lignin, can be seen in the changes in the C(1s) and O(1s) XPS spectra. Figure 92 and 93 show the XPS spectra for the conversion of lignin during the carbon fiber production process.

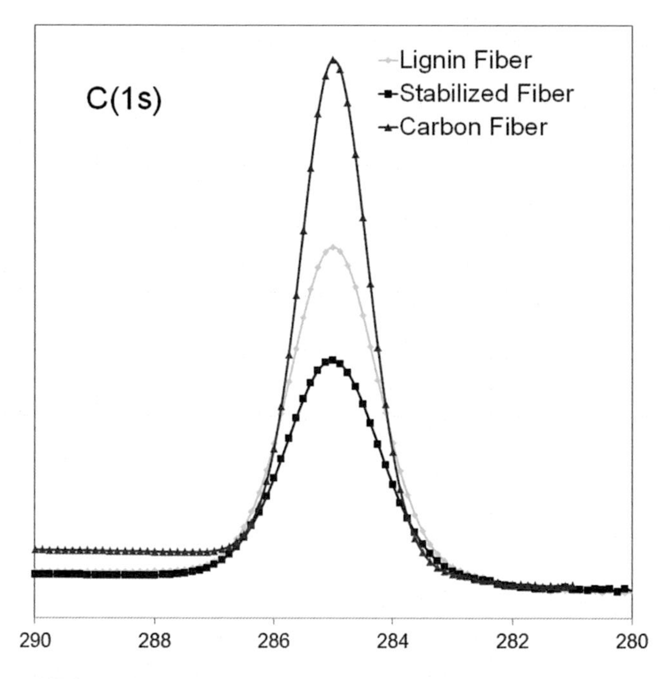

Figure 92: XPS analysis C(1s) of lignin fiber, stabilized lignin fiber and carbonized lignin fiber during conversion process

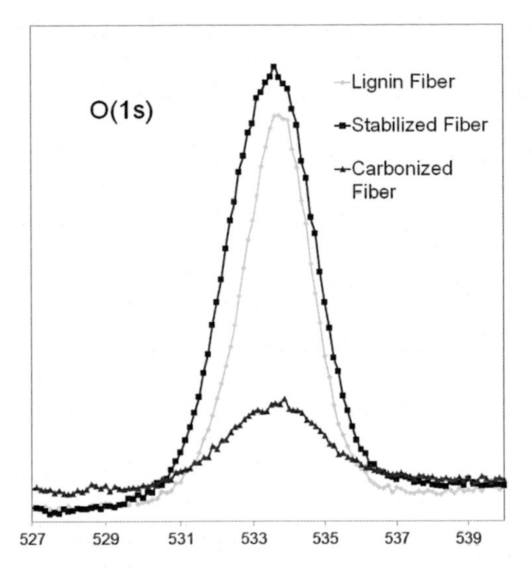

Figure 93: XPS analysis O(1s) of lignin fiber, stabilized lignin fiber and carbonized lignin fiber during conversion process

In the lignin fiber the relation of 0.24 between oxygen and carbon is a little bit below the theoretical value of 0.33 [7-33]. This may be due to the possibility of carbon contaminations such as organic agents and air.

The C/O relation of the oxidized lignin fiber is around 70% higher than the lignin fiber.

The carbonized lignin fiber has the lowest O/C relation as expected.

The low separation of the carbon signals, provide difficulties in obtaining detailed scans to make a quantitative evaluation. In contrast it is easy to see qualitatively the higher C-O components in the oxidized lignin fiber (e.g. C-O, O-C=O). The high content of C-C bonding in the carbonized lignin fiber is also visible.

7.3 Correlations between Properties and Chemical Structure of Lignin Based Carbon Fiber

This chapter showed the improvement of the mechanical properties and the chemical structure of the lignin fiber during conversion process and how they correlate.

The improvement of the mechanical properties was detected by performing the single fiber tensile test. The density of the fibers was measured in a helium pycnometer, and the SEM made the structure of the fiber visible.

The improvement of the chemical structure during conversion of the fiber was detected by performing Raman Spectroscopy and XPS.

The mechanical properties improved during the conversion process of lignin to carbon fiber (Compare Single Fiber Tensile Test – Chapter 7.1.1).

Table 26: Mechanical properties during conversion of Lignin to carbon fiber

	Peak stress [MPa]	**Modulus [Gpa]**
Stabilized Lignin Fiber	29.92	2.28
Lignin based Carbon Fiber	151	20.27

During the conversion process also the density rises. The reason for this is the rising of the carbon contend in the fiber and a more compact microstructure (Compare Chapter 7.1.2 and 7.2.2 as well as Table 27 and 28).

Table 27: Densities of the fibers during conversion process

	Average Density [g/cm³]
Lignin Fiber	1.2921
Stabilized Lignin Fiber	1.5188
Lignin based Carbon Fiber	1.9064

The XPS measurements were performed to show the change of the composition of the fiber during conversion process. As expected the oxygen content of the fiber rises during the stabilization step, and the lignin based carbon fiber has the highest carbon content. The detailed XPS scans and the improvement of the carbon-carbon-bonding can be found in Chapter 7.2.2.

Table 28: Detected elements by XPS during conversion

Atom %	**Carbon**	**Oxygen**
Lignin Fiber	80.5	19.5
Stabilized Lignin Fiber	72.1	27.9
Lignin based Carbon Fiber	94.2	5.8

The rising of the mechanical properties is connected to the structural changes in the fiber during conversion process. For detecting this changes Raman spectroscopy was a very useful tool. It clearly shows the change from a polymer structure (Lignin Fiber) to a highly organized structure of graphitic and diamond bonded carbon structure (Lignin based Carbon Fiber) in Figure 94.

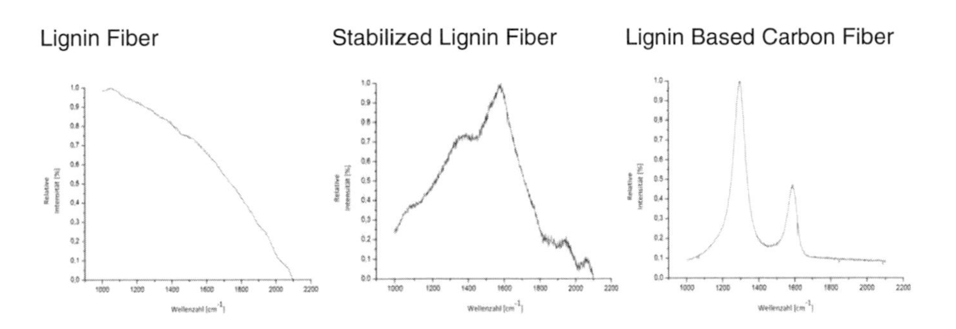

Figure 94: Raman spectroscopy during conversion process

The chemical structure would make much higher mechanical properties possible. Unfortunately, as shown above in the SEM images, the macroscopic defects and pores works against the good chemical structure of the Lignin based Carbon Fiber. (Compare SEM in Chapter 7.1.3)

7.4 References

[7-1] DIN EN ISO 527-1, Bestimmung der Zugeigenschaften, 1996

[7-2] U. Fischer, Tabellenbuch Metall, Verlag Europa Lehrmittel, 2008

[7-3] Din EN ISO 291 Kunststoffe - Normalklimate für Konditionierung und Prüfung (ISO 291:2008); Deutsche Fassung EN ISO 291:2008

[7-4] DIN EN ISO 11566 Kohlenstoffasern - Bestimmung der Zugeigenschaften von Probekörpern aus Einzelfilamenten (ISO 11566: 1996) Deutsche Fassung EN ISO 11566:1996

[7-5] DIN EN ISO 11566 Kohlenstoffasern - Bestimmung der Zugeigenschaften von Probekörpern aus Einzelfilamenten (ISO 11566: 1996) Deutsche Fassung EN ISO 11566:1996

[7-6] DIN EN ISO 1973 Textilien - Fasern - Bestimmung der Feinheit - Gravimetrisches Verfahren und Schwingungsverfahren (ISO 1973:1995); Deutsche Fassung EN ISO 1973:1995

[7-7] Draper, John William (1861). A Textbook on chemistry. p. 46.

[7-8] U.S. Patent 2,667,782 Shea, "Apparatus for measuring volumes of solid materials".

[7-9] U.S. Patent 4,083,228 Turner et al., "Gas comparison pycnometer".

[7-10] U.S. Patent 4,888,718 Furuse, "Volume measuring apparatus and method"

[7-11] J. Eichler, Physik für das Ingenieurstudium, Vieweg + Teubner Verlag, 2011

[7-12] S. B. Hornbogen, Mikro- und Nanoskopie der Werkstoffe, Springer Verlag, 2009

[7-13] G. Fasching, Werkstoffe für die Elektrotechnik: Mikrophysik, Struktur, Eigenschaften, Springer Verlag, 2005

[7-14] Carl-Zeiss GmbH, Handbuch: Supra 55 VP, 2004

[7-15] V. Brückner, Elemente optische Netze. Grundlagen und Praxis der optischen Datenübertragung, Vieweg + Teubner Verlag, 2011

[7-16] W. Demtröder, Laserspektroskopie 1, Springer Verlag, 2011

[7-17] J. Eichler, Physik für das Ingenieurstudium, Vieweg + Teubner Verlag, 2011

[7-18] Z. B. Hesse, Spektroskopische Methoden in der organischen Chemie, Georg Thieme Verlag, 2005

[7-19] S. B. Hornbogen, Mikro- und Nanoskopie der Werkstoffe, Springer Verlag, 2009

[7-20] H.-J. Bargel, Werkstoffkunde, Berlin: Springer Verlag, 2008

[7-21] H. Emminger, Physikum, Stuttgart: Thieme Verlag, 2003

[7-22] H.P. Latscha, Anorganische Chemie, Chemie - Basiswissen, Berlin: Springer, 1992

[7-23] J.G. Graselli, Analytical Raman spectroscopy, John Wiley & Sons Inc, 1991

[7-24] T.J. Dines, Carbon, 1991

[7-25] L. Tuinstra, Raman spectrum of graphite, The Journal of Chemical Physics, 53, 1970, 3

[7-26] J. J. Yeh, I. Lindau: Atomic subshell photoionization cross sections and asymmetry parameters; Atomic data and nuclear data tables. 32, 1985, 1 - 155

[7-27] D. Briggs, M. P. Seah: „Practical Surface Analysis, Vol. 1“, Wiley, 1990, MaWi-Bibliothek (Nr.668)

[7-28] G. Ertl, J. Küppers: „Low Energy Electrons and Surface Chemistry“, Wiley-VCH, 1985, (Nr. 962)

[7-29] J. C. Riviere: „Surface Analytical Techniques“, Clarendon Press Oxford, 1990, MaWi-Bibliothek (Nr. 494)

[7-30] W. Göpel, C. Ziegler: „Struktur der Materie“, Teubner, Leibniz 1994

[7-31] J. Chastain, R. C. King (ed.): „Handbook of X-Ray Photoelectron Spectroscopy“, Physical Electronics, im Fachgebiet vorhanden

[7-32] M. Grasserbauer, H. J. Dudek, M. F. Ebel: „Angewandte Oberflächenanalyse“, Spinger, 1985, MaWi-Bibliothek (Nr. 299)

[7-33] L.-S. Johanson et al., Applied Surface Science 144, 92 (1992)

8 Lignin and Conventional Carbon Fiber

To ensure, that lignin is a precursor with high potential for carbon fiber production a comparison with conventional PAN-based carbon fiber is made in this chapter. Raman and XPS studies show advantages of the chemical structure of the Lignin based Carbon Fiber. REM images demonstrate the challenges of Lignin based carbon Fiber in comparison to conventional PAN-based carbon fiber in this chapter.

8.1 Raman Spectroscopy

Raman spectroscopy makes it possible to compare the chemical structure of the conventional PAN-based carbon fiber and the lignin based carbon fiber. The ability of the Raman spectroscopy to illustrate the different kinds of bonding of carbon is fundamental for this comparison. The modification of the carbon atoms in the fiber is measured. In the Raman spectrum this modifications are displayed by the peaks of the D and G band. The D band represents the sp^2 modification of the carbon and the G band the sp^3 modification. The sharpness of the D and G band in the spectrum is also the indicator of the order of the carbon atoms in the fiber and the crystallinity.

The theoretical background and the experimental conditions of the Raman spectroscopy were presented in Chapter 7.2.1. For the comparison of the Lignin based Carbon Fiber and the conventional carbon fiber, the same instruments and same conditions were used.

The result of the comparison of conventional carbon fiber and lignin based carbon fiber is shown in Figure 95.

The conventional carbon fiber has by far weaker D and G bands (sp^3 and sp^2 hybridization of the carbon) than the Lignin based Carbon Fiber. This is a sign for a higher degree of disorder in the conventional carbon fiber. The lignin based carbon fiber clearly shows sharp D and G bands, which is an indication of a higher order and crystallinity in the fiber.

A higher order and crystallinity leads to higher mechanical properties like tensile strength and Young's modulus. But a higher order and crystallinity also leads to more brittleness. That means the Lignin based Carbon fiber reacts very vulnerable against stress.

It is possible, that foreign atoms like nitrogen in the conventional carbon fiber leads to lower order and crystallinity, which makes the conventional fiber less brittle and less vulnerable.

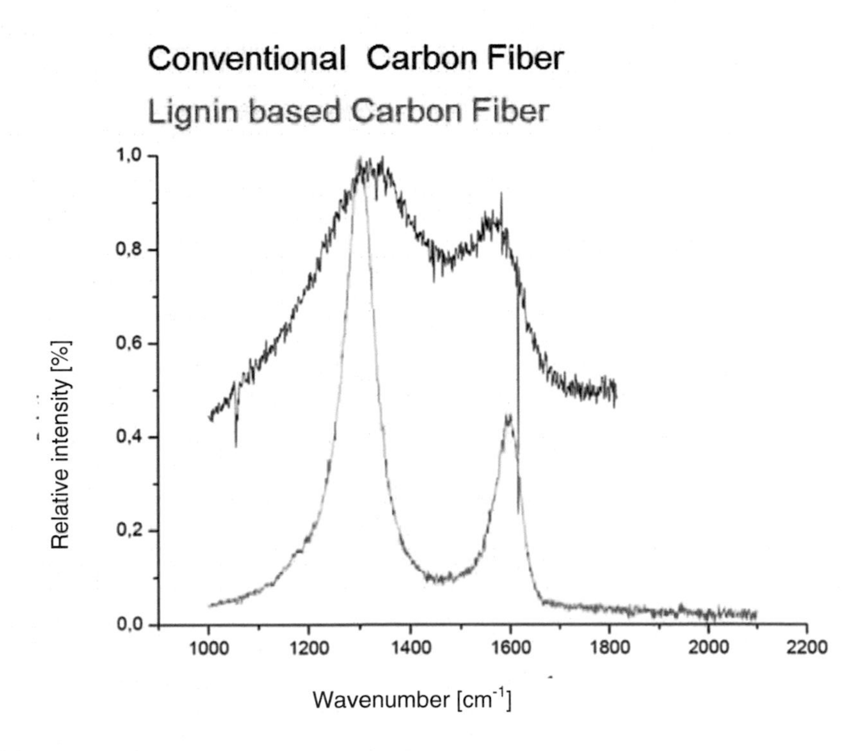

Figure 95: Raman Spectroscopy: Comparison of Lignin based Carbon Fiber and conventional carbon fiber

The results of the Raman spectroscopy leads to the assumption, that the Lignin based Carbon Fiber have the potential for very high mechanical properties, which could be also much higher than the mechanical properties of the conventional carbon fiber.

8.2 XPS Studies

XPS has the ability to detect the atoms and thier bonding, which are contained in the carbon fiber. A comparison of the overview spectrums and high resolution carbon spectrums makes it possible to compare the chemical structures of a lignin-based and a conventional PAN-based carbon fiber.

The theoretical background and the experimental conditions of the XPS studies were presented in Chapter 7.2.1. For the comparison of the Lignin based Carbon Fiber and the conventional carbon fiber, the same instruments and same conditions were used.

The result of the comparison of conventional carbon fiber and lignin based carbon fiber is shown in Table 29, 30 and 31.

For lignin-based carbon fiber and the conventional PAN-based carbon fiber overview spectrums and high resolution carbon spectrums were performed. From the overview spectrums the composition of the fibers was collected. The lignin-based carbon fiber has higher carbon content than the conventional PAN-based carbon fiber. The oxygen content of the lignin based carbon fiber is lower and the lignin based carbon fiber has no nitrogen content.

Table 29: Elements detected by XPS

	Carbon	**Oxygen**	**Nitrogen**
Lignin based Carbon Fiber	94.2	5.8	0.0
Conventional Carbon fiber	87.0	8.0	5.0

Based on the overview spectrums the relation O/C was calculated. The following table presents the relation of O/C.

Table 30: Detection of the O/C Relation by XPS

	Lignin based Carbon Fiber	**Conventional Carbon Fiber**
O/C	0.06	0.09

The high resolution spectrums show the content of the aliphatic carbon-carbon and carbon-carbon-donble-bonds. The determined high resolution spectrums have an absolute error, which is smaller than ± 0.03.

For the elements oxygen and carbon, detailed high resolution scans were made. The carbon signal was fitted with five groups of components:

- C-C, C-H, C=C (285.0 eV)
- C-O, C-C-O (286,5 eV)
- C=O, O-C-O (288.0 eV)
- O-C=C (289.0 eV)
- shake up* (291.5 eV)

Since the oxygen signal was less sharp, it was fitted with one group of components. The resulting bonding-energy (E_B) and the fractions of bonding (%) are shown in Table 26. The fitting interval was ⩽0.5 eV.

Table 31: Detailed XPS scans of oxygen and carbon

Region	**Class**	**E_B/eV**	**Conventional Carbon Fiber [%]**	**Lignin based Carbon Fiber [%]**
C1s	C-C	285.0	60.4	66.0
	C-H	285.0	60.4	66.0
	C=C	285.0	60.4	66.0
	C-O	286.5	14.1	14.3
	C-O-C	286.5	14.1	14.3

C1s	C=O	288.0	7.8	4.1
	O-C-O	288.0	7.8	4.1
	O-C=O	289.0	6.0	7.0
	Shake up*	291.5	2.5	2.4
O1s	O-C	533.7	9.2	6.2
	C-O-C	533.7	9.2	6.2
	C=O	533.7	9.2	6.2
	O-C-O	533.7	9.2	6.2
	O-C=O	533.7	9.2	6.2

* "Shake up" Carbon in aromatic compound (crossing $\pi - \pi^*$)

The conventional carbon fiber has lower carbon content than the lignin based carbon fiber. The carbon content of the lignin based carbon fiber is more than 7% higher than conventional carbon fiber. Also the C1s signal of the carbon-carbon bond and the carbon-carbon double bonnding of the lignin based carbon fiber is stronger than the ones of the conventional carbon fiber. The conventional carbon fiber has higher oxygen content and thus more carbon-oxygen bonding.

Since more carbon-carbon bonding as well as carbon-carbon double bonding leads to higher mechanical properties in the fiber the results of XPS leads to the assumption, that the lignin based carbon fiber has the potential for very high mechanical properties. These mechanical properties could be much higher than the mechanical properties of the conventional carbon fiber.

The results of the Raman Spectroscopy and the XPS fit qualitative together and show the high potential of lignin based carbon fiber.

8.3 Scanning Electron Microscopy

The results of the Raman Spectroscopy and the XPS fit qualitative together and show the high potential of lignin based carbon fiber by characterization of the chemical structure. Unfortunately the macroscopic defects and pores in the lignin based carbon fiber works against the good chemical structure. Scanning Electron Microscopy is a useful tool to show these effects.

The theoretical background as well the experimental conditions of the Scanning Electron Microscopy studies were presented in Chapter 7.1.3. For the comparison of the lignin based carbon fiber and the conventional carbon fiber, the same instruments and same conditions were used.

The Scanning Electron Microscopy (SEM) images clearly show the actual problems and the reason, why high mechanical properties for the lignin based carbon fibers are not reachable right now. The pores and defects reduce the area cross-section and induct superposition of stress in the fiber. Both effects leads to lower mechanical properties of the lignin based carbon fiber.

Figure 83 compares the lignin based carbon fiber (Figure 96 a, b, c, d) and conventional carbon fiber (Figure 96 e, f, g, h).

Lignin based Carbon Fiber

a) overview image

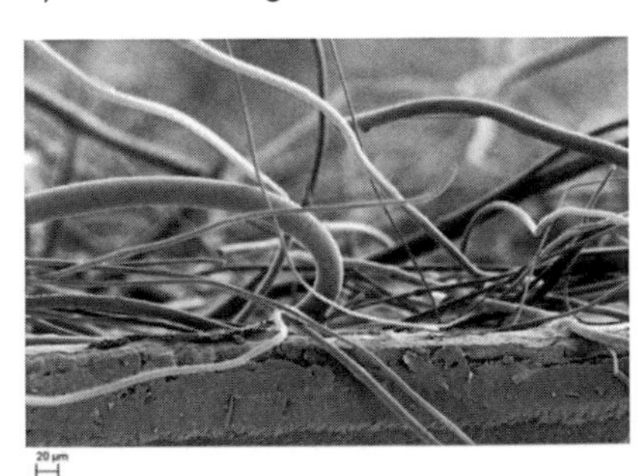

Conventional Carbon Fiber

e) overview image

b) defects at the surface

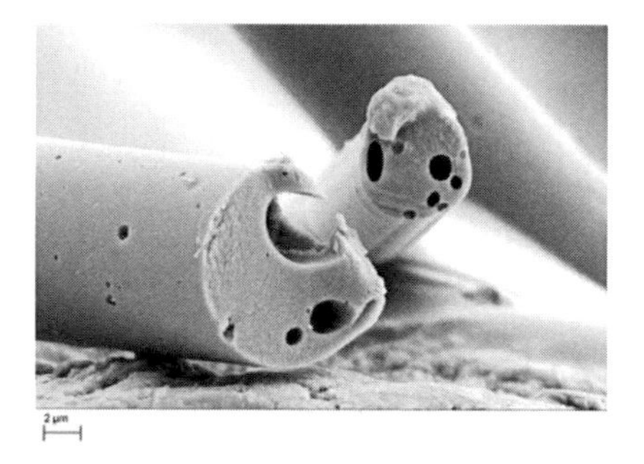

c) defects at the cross section

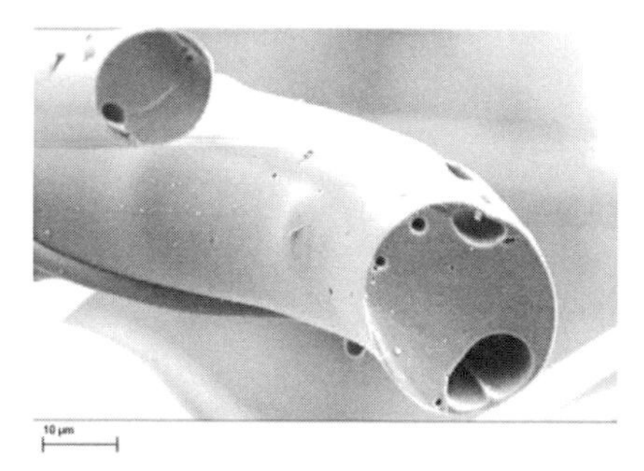

d) defects and inhomogeneity

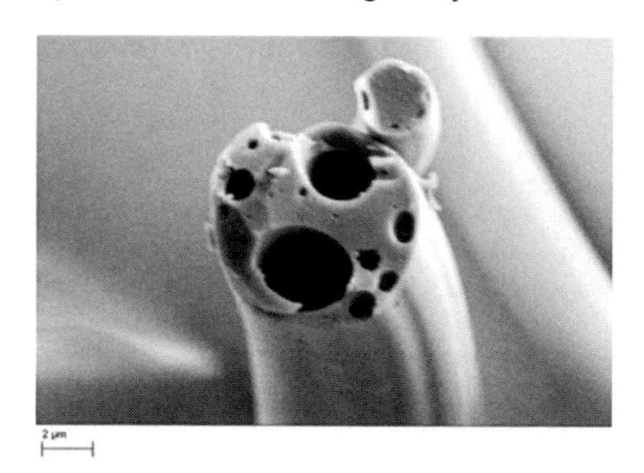

f) cross section

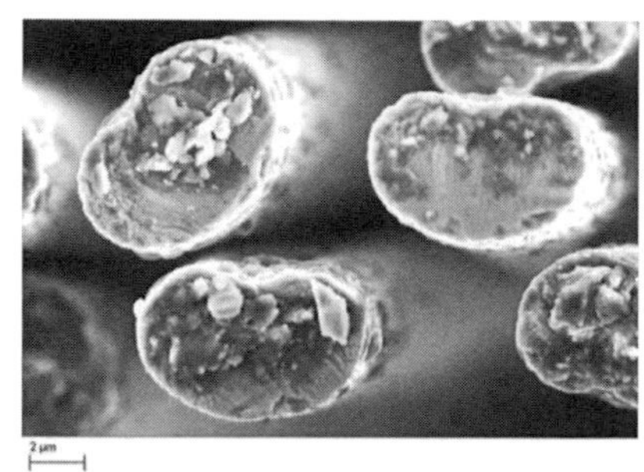

g) *surface and cross section*

h) *homogeneity*

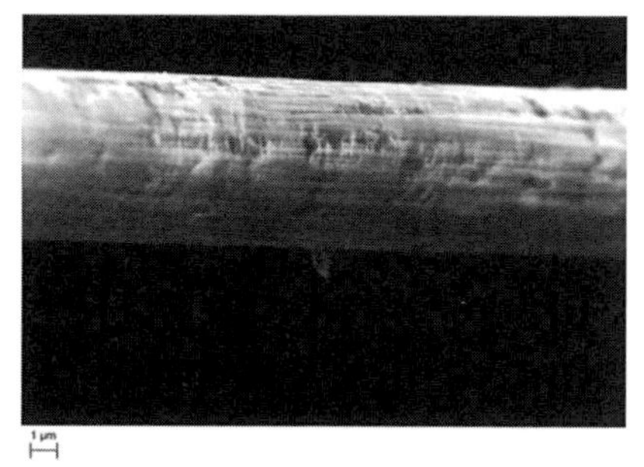

Figure 96: a, b, c, d SEM images of lignin based carbon fiber and e, f, g, h SEM images of conventional carbon fiber

The macroscopic defects (compare Figure 96 a, b, c, d) in the lignin based carbon fiber are responsible for the lower mechanical properties in comparison to conventional carbon fiber. The strength and stiffness of a conventional carbon fiber is more than ten times higher than the mechanical properties of the lignin based carbon fiber. The better mechanical properties of the conventional carbon fiber are also connected to the size effect. The size effect is not working for the lignin based carbon fibers since the defects stays at the same size when the fiber is shrinking during conversion process. The defects lead to stress concentration and that's the reason for the lower mechanical properties of the lignin based carbon fiber. The better chemical structure of the Lignin based Carbon Fiber is not able to compensate this effect. Since in the conventional carbon fiber no stress concentration is visible, the mechanical properties are not influenced negatively by this effect. The higher order of the carbon atoms and the higher degree of crystallinity of the lignin based carbon fiber makes the fiber stiffer but also more brittle.

The mechanical Properties of the lignin based carbon fiber could be much better, if it's possible to reduce the macroscopic defects. The chemical structure leads to the assumption, that the lignin based carbon fiber has the potential for very high mechanical properties. In summary the Chapter 7 and 8 show the high potential of lignin based carbon fiber.

9 Potential of Lignin-Based Carbon Fiber

The following chapter shall give an overview about the potential of the use of lignin based carbon fiber for automotive applications in the near future and in the long term.

9.1 Economic and Ecological Potential of Lignin-Based Carbon Fiber

The potential of lignin based carbon fiber is directly connected to the chemical structure of the fiber. Chapter 7 and 8 clearly shows the potential of the lignin based carbon fiber by using several analytical techniques. It turns out, that the chemical structure is even better than the chemical structure of conventional carbon fiber. For these reason higher mechanical properties in the long term seems to be possible.

Also the possibility of using conventional conversion lines in the carbon fiber industry lowers investment cost dramatically. Existing carbon fiber production lines can be used for this new carbon fiber. This and the low price of Hardwood Lignin lead to lower carbon fiber prices. The result of the calculation of carbon fiber prices is shown in Figure 84: Economical potential of alternative precursors. Cost reductions of the full carbon fiber price of over 40% are possible. Also enormous savings of CO_2 emissions are possible by using Hardwood Lignin as an alternative precursor. Since Hardwood Lignin is a sustainable and renewable resource and available as a waste product of the paper mill industry and the bio refinery CO_2 emissions can be saved by up to 60% compared with the conventional carbon fiber.

The chemical structure as well as the economic and ecological potential of the lignin based carbon fiber summarizes perfectly the capability of this alternative precursor for carbon fiber production. This thesis shows how a sustainable, low cost automotive carbon fiber can become available. With rising properties the lignin based carbon fiber will be used for structural applications in the automotive industry.

Chapter 9.2 will now show examples of applications which will be available with today's mechanical properties.

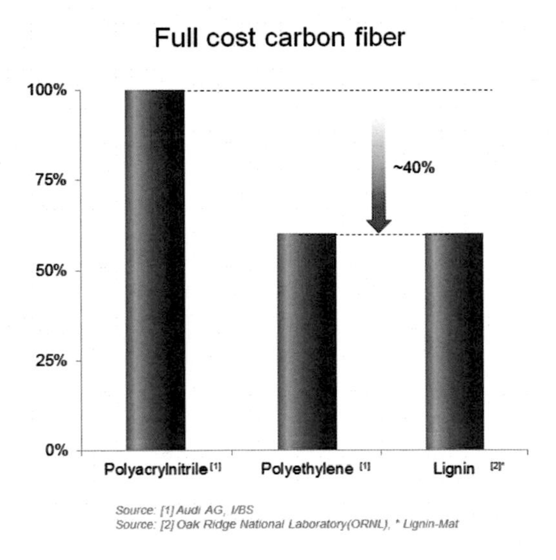

Figure 97: Economical potential of alternative precursors [9-1, 9-2]

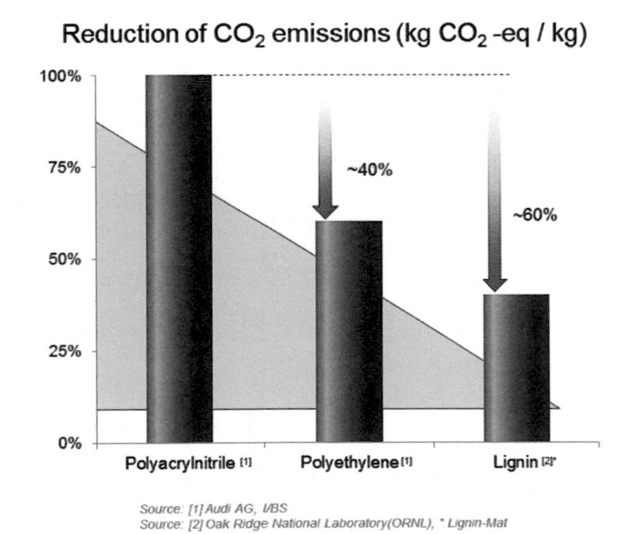

Figure 98: Ecological potential of alternative precursors [9-1, 9-2]

9.2 Possible products made from Lignin based Carbon Fiber in the near future

There are several propertles besides the mechanical properties which make it possible to use lignin in automotive applications. Figure 99 gives an overview concerning properties and possible products made from Lignin based Carbon Fiber.

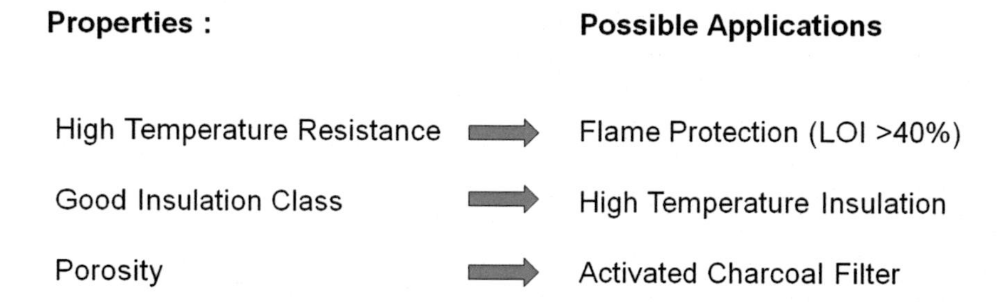

Figure 99: Possible Applications for Lignin based Carbon Fiber

The high LOI number (Limiting Oxygen Index) of over 40% and the form of a flexible mate makes Lignin based Carbon Fiber a perfect material for flame protection. A concentration of oxygen of over 40% is necessary to ignite the material. Since air has oxygen concentration of 19% the material is practically inflammable. A possible application could be the firewall in cars, which separates the engine compartment from the cabin.

The flame resistance in connection with the high insulation class also makes it possible to think about high temperature insulation in cars. The major thermal properties of Lignin based Carbon Fiber were measured, and are shown in Table 32. Interesting for automotive applications are especially the good thermal properties at high temperatures, and the low density of this material. The engine compartment or exhaust gas system are possible areas where this high temperature insulation could work perfectly.

Table 32: Thermal properties of Lignin based Carbon Fiber

Characteristic	Unit	In Fiber direction	Against Fiber Direction
Bulk Density	g/cm³	0,18	0,18
Apparent Porosity	%	90	90
Thermal Conductivity 25°C	W/mK	0.39	0.18
Thermal Conductivity 1500°C	W/mK	1.5	0.45
α (RT to 1000°C)	10^{-6}/°C	3.0	3.3
α (RT to 2000°C)	10^{-6}/°C	3.6	4.0

Besides the thermal properties also the porosity of Lignin based Carbon fiber offers opportunities in the automotive industry. The use of the Lignin based Carbon Fiber mate as an active charcoal filter offers a filtration system for continuous removement of CO_2, H_2O and hazardous contaminants and odors from cabin air. The concept is presented in Figure 100.

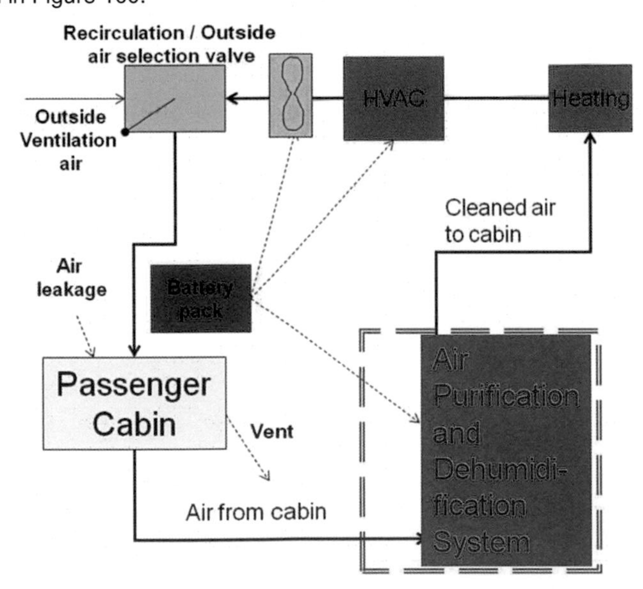

Figure 100: Air filter using Lignin based Carbon Fiber

The active charcoal filter made from Lignin based carbon fiber allows a safe operation of the vehicle with limited outdoor air ventilation, which causes energy saving. Up to 40% energy savings can be achieved by replacing outside dilution air by purified and dehumidified recycled air. That means extending of the driving range of EVs by about 10% and saving of 10% of fuel for ICE vehicles.

This three examples show how Lignin based Carbon Fiber could already change the automotive industry today.

9.3 References

[9-1] C. Eberle, Oak Ridge Carbon Fiber Composites Consortium Meeting 7th to 8th of April 2012, Oak Ridge: Lignin-based Precursors

[9-2] C. Eberle, T. Albers, C. Chen ,D. Webb Commercialization of New Carbon Fiber Materials Based on Sustainable Resources for Energy Applications (ORNL/TM-2013/54) Published: March 2013

10 General Conclusions

The goal of this dissertation is to explain the research to produce a lignin based carbon fiber.

This dissertation shows for the first time in research the complete production process of carbon fiber made from lignin by analyzing the chemical structure of all intermediate products, and determines the potential of lignin as an alternative, economic and ecological precursor.

Currently the drawbacks of using CFRPs are the high manufacturing cost. This is due to the fact that over 50% of the cost of a PAN based carbon fiber, belongs to the manufacturing cost of the precursor. Accordingly the manufacturing cost of carbon fiber can be significantly reduced by the use of alternative precursors. The decision to use lignin as an alternative precursor to produce carbon fiber was based on the low price, the good availability, and the renewability of the precursor in question.

To investigate the potential of lignin as an alternative precursor, a set of fundamental production steps were established, since there was no procedure to produce carbon fiber from lignin. These steps consist of: pelletizing, fiber spinning, fiber stabilization, and carbonization. The intermediate products of these steps are the lignin powder, the lignin pellets, the lignin fiber, as well as the oxidized and carbonized lignin fiber. At each step the properties and chemical structure were analyzed.

This chapter summarizes the findings of this project and highlights some recommendations of future work.

10.1 Development of lignin based carbon fiber

The production process of a PAN based carbon fiber contains fiber spinning, fiber stabilization, and fiber carbonization as the major steps. The lignin based carbon fiber production process is comparable to the PAN based carbon fiber production process, but two steps were added. These two steps are washing and drying of lignin powder as the first step, and pelletizing of the lignin powder as the second step. This leads to five main steps:

- Washing and drying of lignin powder
- Pelletizing of the lignin powder
- Melt spinning of the lignin fiber
- Oxidation of the fiber
- Carbonization of the fiber

10.2 Proof of industrial production size

Today the demand on carbon fiber is increasing. A potential alternative precursor must be capable of mass production. This dissertation demonstrates large scale production of the lignin based carbon fiber.

The large scale pelletizing process was done with a 53 mm diameter extrusion machine, which was equipped with a hot die face cutter. 1000 kg of Hardwood Lignin was successfully pelletized for subsequent melt spinning into Hardwood Lignin precursor fiber.

Using a semi production scale machine, “melt blown” spinning was performed for producing a Hardwood Lignin fiber web, with a filament diameter ranged from 10 to 20 µm. At rates approaching 15 kg/h, the fibers were spun into a web approximately 60cm wide with an areal density of 230 g/m^2.

Using the Hardwood Lignin fiber web, stabilized fibers were produced in a batch thermal treatment process. The fibers were placed in a large (> $5.5m^3$) oven. In one batch 75kg of lignin fibers were stabilized for further processing.

In the laboratory, the stabilized fibers were heat-treated to produce ~ 25 kg of lignin based carbon fibers in one batch. The carbonization process can occur at temperatures ranging from 500 to 1500 °C for 5 to 10 min. Approximately 65% of the material is vaporized during carbonization, with gasses exhausted through an incineration system.

This is the first reported production of lignin based carbon fibers at a scale exceeding 1 kg and shows the possibility to scale up the production process in an industrial size.

10.3 Properties and chemical characterization

To qualify lignin as a precursor, a detailed characterization of the properties and the chemical structure of the lignin powder was necessary.

The thermal properties were obtained by thermo gravimetric analysis and differential scanning calorimetry.

The chemical structure was defined using the following techniques:

1. Elementary analysis for detection of the C9-units
2. mass spectroscopy for analysis of defragmentation and side-chains
3. nuclear magnetic resonance spectroscopy and Fourier transform infrared spectroscopy for complete analysis of chemical bonding in the lignin macro-molecule and defining of the chemical structure

On the basic of these experiments the chemical structural formula for the used Hardwood Lignin was proposed, which is necessary for the understanding of the chemical reactions during conversion of lignin to carbon fiber. The major chemical reactions of the conversion process were verified, using the same analytical techniques as mentioned above.

The improvement of the mechanical properties was detected by performing the single fiber tensile test. The density of the fibers was measured in a helium pycnometer, and the SEM made the structure of the fiber visible.

The improvement of the chemical structure during conversion of the fiber was detected by performing Raman Spectroscopy and XPS. The improvement of the mechanical properties is connected to the structural changes in the fiber during conversion process. For detecting this structural changes Raman spectroscopy was a very useful tool. It clearly shows the change from a polymer structure (Lignin Fiber) to a highly organized structure of graphitic and diamond bonded carbon structure (Lignin based Carbon Fiber).

The chemical structure would make much higher mechanical properties possible. Unfortunately, as shown in the SEM images, the macroscopic defects, such as pores works against the good chemical structure of the lignin based carbon fiber.

10.4 Economic and ecological potential of lignin based carbon fiber

Existing carbon fiber production lines can be used for this new precursor. This possibility lowers investment cost dramatically. In connection with the low price of Hardwood Lignin, this leads to lower carbon fiber prices. The result of the calculation of carbon fiber prices shows the economic potential of alternative precursors. Cost reductions of the full carbon fiber price of over 40% are possible. But not only economic reasons make lignin attractive as a precursor, also ecological reasons are important. Enormous savings of CO_2 emissions are possible by using Hardwood Lignin as an alternative precursor. Since Hardwood Lignin is a sustainable and renewable resource, and available as a waste product of the paper mill industry and the bio refinery, CO_2 emissions can be saved by up to 60% compared with the PAN based carbon fiber production.

The chemical structure as well as the economic and ecological potential of the lignin based carbon fiber summarizes perfectly the capability of this alternative precursor for carbon fiber production. This thesis shows how a sustainable, low cost automotive carbon fiber can be achieved. With higher properties the lignin based carbon fiber will be used for structural applications in the automotive industry.

10.5 Recommendations for future work

Since the chemical structure of lignin based carbon fiber showed the potential of this precursor, the process technology of the fiber spinning process will be the focus of future research and development. A reduction of defects, such as pores, in the fiber will lead to much better mechanical properties of the lignin based carbon fiber.

The chemical modifications and purification of the lignin will also be a subject of future research. The removal of volatile compounds will help to perform the melt spinning process of the lignin fiber without the implementation of pores in the fiber. This will also lead to much higher mechanical properties.

11 List of Publication

I. Reviewed Paper + Presentation

Society of Plastics Engineers (Automotive Division) - Automotive Composites Conference & Exhibition 2013, Novi, Mich. USA

Hendrik Mainka[1], Oliver Stoll[1], Enrico Koerner[1], Olaf Taeger[1]

[1]Volkswagen AG, Berliner Ring 2, Wolfsburg (Germany)

Axel S. Herrmann[2]

[2]Faserinstitut Bremen e.V., Am Biologischen Graten 2,
28359 Bremen (Germany)

Alternative Precursors for sustainable and cost-effective carbon fibers usable within the automotive industry

II. Presentation

2013 Composites World Carbon Fiber Conference, Knoxville, TN USA

Hendrik Mainka[1], Oliver Stoll[1], Enrico Koerner[1], Olaf Taeger[1]

[1]Volkswagen AG, Berliner Ring 2, Wolfsburg (Germany)

Axel S. Herrmann[2]

[2]Faserinstitut Bremen e.V., Am Biologischen Graten 2,
28359 Bremen (Germany)

Sustainable and cost-effective carbon fibers made from lignin usable within the automotive industry

III. Reviewed Paper + Presentation

Society of Plastics Engineers (Automotive Division) - Automotive Composites Conference & Exhibition 2014, Novi, Mich. USA

Hendrik Mainka[1], Oliver Stoll[1], Enrico Koerner[1], Olaf Taeger[1]

[1]Volkswagen AG, Berliner Ring 2, Wolfsburg (Germany)

Axel S. Herrmann[2]

[2]Faserinstitut Bremen e.V., Am Biologischen Graten 2,
28359 Bremen (Germany)

Chemical Characterization of Lignin as a Precursor for Automotive Carbon Fiber

IV. Reviewed Paper

ATZ Automobiltechnische Zeitschrift (24.02.2015 - Volume 03/2015, pp 52-57 – DOI: 10.1007/s35148-015-0403-0)

Hendrik Mainka[1], Oliver Stoll[1], Enrico Koerner[1], Olaf Taeger[1]

[1]Volkswagen AG, Berliner Ring 2, Wolfsburg (Germany)

Axel S. Herrmann[2]

[2]Faserinstitut Bremen e.V., Am Biologischen Graten 2,
28359 Bremen (Germany)

Maria Laue[3], Rolf-Dieter Hund[3]

[3]ITM der TU Dresden, Zellescher Weg 19, 01062 Dresden (Germany)

Ligninpulver als Füllstoff für thermoplastische Leichtbaukomponenten

V. Reviewed Paper

ATZ Automobiltechnische Zeitschrift (24.02.2015 Volume 117, Issue 3 , pp 32-35 – DOI: 10.1007/s38311-015-0171-1)

Hendrik Mainka[1], Oliver Stoll[1], Enrico Koerner[1], Olaf Taeger[1]

[1]Volkswagen AG, Berliner Ring 2, Wolfsburg (Germany)

Axel S. Herrmann[2]

[2]Faserinstitut Bremen e.V., Am Biologischen Graten 2,

28359 Bremen (Germany)

Maria Laue[3], Rolf-Dieter Hund[3]

[3]ITM der TU Dresden, Zellescher Weg 19, 01062 Dresden (Germany)

Lignin: A Filler for Thermoplastic Automotive Lightweight Components

VI. Reviewed Paper + Presentation

Society of Plastics Engineers (Automotive Division) - Automotive Composites Conference & Exhibition 2015, Novi, Mich. USA

Hendrik Mainka[1],

[1]Volkswagen AG, Berliner Ring 2, Wolfsburg (Germany)

Axel S. Herrmann[2]

[2]Faserinstitut Bremen e.V., Am Biologischen Graten 2,

28359 Bremen (Germany)

Raman and X-ray Photoelectron Spectroscopy: useful tools for the chemical characterization of the conversion process of Lignin to carbon fiber

VII. Patent Pending

Schutzrechtsanmeldung - Aktenzeichen 10 2014 215 627.0

Hendrik Mainka[1], Christine Schütz[1], Jörg Hain[1]

[1]Volkswagen AG, Berliner Ring 2, Wolfsburg (Germany)

Verbundwerkstoff sowie Formteile aus diesem

VIII. Reviewed Paper

Journal of Materials Research and Technology (Mar 31, 2015 - DOI Information: 10.1016/j.jmrt.2015.03.004)

Hendrik Mainka[1],

[1]Volkswagen AG, Berliner Ring 2, Wolfsburg (Germany)

Axel S. Herrmann[2]

[2]Faserinstitut Bremen e.V., Am Biologischen Graten 2,
28359 Bremen (Germany)

LIGNIN - AN ALTERNATIVE PRECURSOR FOR SUSTAINABLE AND COST-EFFECTIVE AUTOMOTIVE CARBON FIBER

IX. Reviewed Paper

Journal of Materials Research and Technology (Apr 14, 2015 - DOI Information: 10.1016/j.jmrt.2015.04.005)

Hendrik Mainka[1],

[1]Volkswagen AG, Berliner Ring 2, Wolfsburg (Germany)

Axel S. Herrmann[2]

[2]Faserinstitut Bremen e.V., Am Biologischen Graten 2,
28359 Bremen (Germany)

Characterization of the major reactions during conversion of lignin to carbon fiber

12 List of Abbreviations and Symbols

A	Area cross-section
ATR	Attenuated total reflection
c	Speed of light
CFK	Kohlenstofffaserverstärkte Kunststoffe
CFRP	Carbon Fiber Reinforced Plastics
-CN	Cyanide group
=C=O	Carboxyl group
CO2	Carbon dioxide
-CO3	Carbonate group
-COOH	Carboxylic acid group
COSY	COrrelated SpectroscopY
CS2	Carbon disulfide
CF	Carbon Fiber
CH2	Methylen group
CH2=CHCN	Acrylonitrile
CH3	Methyl group
(C2H4)nH2	Polyethylene
(C3H3N)n	Polyacrylonitrile
C6H10O5	Cellulose
cm^{-1}	Inverse centimeter
CP/MAS	Solid state NMR
DIN	Deutsche Institut für Normung
DME	dimethyl ether
DMF	N,N-dimetylformamide
DMSO	Dimethylsulfoxide
DSC	Differential Scanning Colorimeter
E	Yong's Modulus
EA	Elementary Analysis
E_B	Binging Energy

$E_K(e^-)$	Kinetic Energy of an Electron
e.g.	For example
EN	Europäische Norm
eV	Electron volt
EPA	Environment Protection Agency
EU	European Union
F	Force
f	Frequency
FTIR	Fourier transform infrared spectroscopy
GPa	Giga Pascal
h	Hour
h	Planck constant
H2O	Water
H2S	Hydrogen sulfide
H2SO4	Sulfuric acid
H4PO4	Phosphoric acid
HSO3-	Hydrogensulfit
HMBC	Heteronuclear Multiple Bond Correlation
HSQC	Heteronuclear Single Quantum Coorrela.
H+	Hydrogen ion
IR	Infrared
ISO	Int. Organization for Standardization
K	Kalvin
k	Wavenumber
kg	Kilogram
kHz	Kilohertz
kV	Kilovolt
kWh	Kilowatt hour
l	Liter
l	Length
LOI	Limiting Oxygen Index
m2	Square meter

m3	Cubic meter
mA	Milliampere
MeO	Methoxy
MeOH	Methanol
mg	Milligram
Min	Minute
ml	Milliliter
mm	Millimeter
mol	Mole
Mol.Wt.	Molecular weight
MPa	Mega Pascal
MS	Mass spectroscopy
MW	Mega Watt
mW	Milliwatt
m/z	Mass-to-charge ratio
NaSCN	Sodium thiocyanate
NaOH	Sodium hydroxide
Na2SO4	Sodium sulfate
N2	Nitrogen
NMR	Nuclear Magnetic Resonance Spectro.
=O	Epoxide
O-CH3	Methoxy group
-OH	Hydroxyl group
PAN	Polyacrylonitrile
PE	Polyethylene
pH	"power of Hydrogen"
PP	Polypropylene
ppm	Parts per million
PVC	Polyvinylchloride
PVD	Physical vapor deposition
RSD	Relative standard deviation
SE	Secondary Electron

SEM	Scanning Electron Microscopy
SH-	Hydrosulfide
SO2	Sulfur dioxide
T	Temperature
T	Yarn count
TCD	Thermal conductivity detector
TG	Glass transmission temperature
TR	Temperature of the reference
TS	Temperature of the sample
TGA	Thermo Gravimetric Analysis
TMDP	Tertamethyl-1,2,3-dioxaphospholane
TMS	Tetramethylsilane
u	Atomic mass unit
XPS	X-ray Photoelectron Spectroscopy
ZnCl2	Zinc chloride
ZrO2	Zirconium dioxide
Δl	Change of the length
E	Elongation
v	Wavelength
ρ	Density
μm	Micrometer
μs	Microsecond
σ	Strength
°	Degree symbol
°C	Degree Celsius
$	Dollar
%	Percent
1D	One-dimensional
1H	Hydrogen isotope
2D	Two-dimensional
13C	Carbon isotope
15N	Nitrogen isotope

31P	Phosphor isotope

13 Appendix

13.1 Results of the Density Measurements

Run Data Lignin Fiber

RUN	VOLUME (cm^3)	DENSITY (g/cm^3)
1	0.3762	1.2920
2	0.3746	1.2975
3	0.3752	1.2956
4	0.3760	1.2927
5	0.3754	1.2949
6	0.3747	1.2974
7	0.3742	1.2991
8	0.3760	1.2927
9	0.3768	1.2902
10	0.3755	1.2946
11	0.3756	1.2941
12	0.3763	1.2916
13	0.3773	1.2885
14	0.3769	1.2896
15	0.3774	1.2879
16	0.3762	1.2921
17	0.3756	1.2941
18	0.3767	1.2903
19	0.3754	1.2948
20	0.3759	1.2930
21	0.3759	1.2930
22	0.3760	1.2927
23	0.3770	1.2892

24	0.3766	1.2909
25	0.3777	1.2869
26	0.3764	1.2915
27	0.3757	1.2938
28	0.3769	1.2896
29	0.3766	1.2908
30	0.3758	1.2935
31	0.3773	1.2883
32	0.3767	1.2903
33	0.3762	1.2923
34	0.3760	1.2929
35	0.3764	1.2913
36	0.3772	1.2887
37	0.3772	1.2886
38	0.3769	1.2899
39	0.3764	1.2914
40	0.3755	1.2947
41	0.3749	1.2966
42	0.3760	1.2930
43	0.3770	1.2892
44	0.3764	1.2916
45	0.3774	1.2882
46	0.3771	1.2889
47	0.3766	1.2907
48	0.3771	1.2891
49	0.3766	1.2908
50	0.3774	1.2882
51	0.3760	1.2930
52	0.3761	1.2925
53	0.3749	1.2965
54	0.3751	1.2958
55	0.3770	1.2894
56	0.3767	1.2905
57	0.3768	1.2900

58	0.3762	1.2920
59	0.3757	1.2938
60	0.3767	1.2905
61	0.3759	1.2933
62	0.3769	1.2896
63	0.3759	1.2931
64	0.3768	1.2901
65	0.3757	1.2937
66	0.3746	1.2976
67	0.3758	1.2936
68	0.3766	1.2909
69	0.3771	1.2891
70	0.3769	1.2897
71	0.3749	1.2966
72	0.3767	1.2906
73	0.3772	1.2887
74	0.3758	1.2936
75	0.3751	1.2958
76	0.3763	1.2919
77	0.3771	1.2890
78	0.3769	1.2896
79	0.3749	1.2966
80	0.3763	1.2919
81	0.3760	1.2928
82	0.3751	1.2960
83	0.3758	1.2933
84	0.3756	1.2941
85	0.3756	1.2940
86	0.3754	1.2948
87	0.3760	1.2928
88	0.3764	1.2915
89	0.3758	1.2934
90	0.3768	1.2902
91	0.3755	1.2944

92	0.3755	1.2946
93	0.3767	1.2903
94	0.3769	1.2898
95	0.3765	1.2911
96	0.3762	1.2922
97	0.3761	1.2924
98	0.3769	1.2898
99	0.3759	1.2932

Run Data Stabilized Lignin Fiber

RUN	VOLUME (cc)	DENSITY (g/cc)
1	0.4007	1.4899
2	0.4010	1.4888
3	0.3990	1.4961
4	0.3998	1.4932
5	0.3988	1.4972
6	0.3968	1.5045
7	0.3952	1.5108
8	0.3962	1.5069
9	0.3947	1.5126
10	0.3957	1.5086
11	0.3949	1.5118
12	0.3939	1.5157
13	0.3939	1.5154
14	0.3951	1.5112
15	0.3925	1.5210
16	0.3924	1.5216
17	0.3931	1.5188
18	0.3941	1.5147
19	0.3929	1.5195

20	0.3930	1.5191
21	0.3930	1.5191
22	0.3946	1.5130
23	0.3934	1.5176
24	0.3931	1.5186
25	0.3924	1.5216
26	0.3927	1.5201
27	0.3933	1.5180
28	0.3935	1.5171
29	0.3934	1.5177
30	0.3936	1.5169
31	0.3925	1.5208
32	0.3917	1.5242
33	0.3931	1.5187
34	0.3941	1.5147
35	0.3920	1.5231
36	0.3922	1.5221
37	0.3915	1.5250
38	0.3925	1.5210
39	0.3935	1.5170
40	0.3929	1.5194
41	0.3934	1.5177
42	0.3928	1.5199
43	0.3937	1.5163
44	0.3919	1.5234
45	0.3920	1.5231
46	0.3932	1.5184
47	0.3924	1.5214
48	0.3918	1.5239
49	0.3923	1.5217
50	0.3928	1.5199
51	0.3937	1.5163
52	0.3920	1.5230
53	0.3910	1.5268

54	0.3919	1.5233
55	0.3926	1.5205
56	0.3931	1.5188
57	0.3931	1.5186
58	0.3927	1.5203
59	0.3926	1.5207
60	0.3920	1.5228
61	0.3923	1.5218
62	0.3923	1.5219
63	0.3919	1.5235
64	0.3922	1.5221
65	0.3924	1.5213
66	0.3927	1.5203
67	0.3924	1.5213
68	0.3929	1.5195
69	0.3916	1.5247
70	0.3918	1.5239
71	0.3919	1.5233
72	0.3921	1.5225
73	0.3920	1.5231
74	0.3916	1.5247
75	0.3909	1.5274
76	0.3919	1.5234
77	0.3920	1.5230
78	0.3918	1.5236
79	0.3920	1.5228
80	0.3921	1.5224
81	0.3916	1.5244
82	0.3924	1.5215
83	0.3923	1.5218
84	0.3926	1.5207
85	0.3914	1.5252
86	0.3923	1.5219
87	0.3916	1.5247

88	0.3929	1.5194
89	0.3922	1.5222
90	0.3917	1.5243
91	0.3904	1.5291
92	0.3921	1.5224
93	0.3920	1.5231
94	0.3931	1.5188
95	0.3919	1.5234
96	0.3912	1.5262
97	0.3922	1.5220
98	0.3919	1.5232
99	0.3935	1.5173

Run Data Lignin based Carbon Fiber

RUN	VOLUME (cc)	DENSITY (g/cc)
1	0.3396	1.8413
2	0.3369	1.8559
3	0.3362	1.8601
4	0.3354	1.8646
5	0.3354	1.8644
6	0.3346	1.8690
7	0.3341	1.8717
8	0.3329	1.8783
9	0.3323	1.8815
10	0.3327	1.8797
11	0.3322	1.8823
12	0.3318	1.8846
13	0.3311	1.8888
14	0.3328	1.8788
15	0.3329	1.8786

16	0.3301	1.8942
17	0.3291	1.8999
18	0.3291	1.9000
19	0.3292	1.8994
20	0.3291	1.9001
21	0.3287	1.9025
22	0.3294	1.8986
23	0.3292	1.8992
24	0.3284	1.9043
25	0.3282	1.9054
26	0.3286	1.9028
27	0.3282	1.9051
28	0.3293	1.8988
29	0.3282	1.9052
30	0.3283	1.9048
31	0.3279	1.9072
32	0.3281	1.9057
33	0.3284	1.9042
34	0.3283	1.9045
35	0.3293	1.8987
36	0.3295	1.8975
37	0.3283	1.9048
38	0.3284	1.9041
39	0.3291	1.8998
40	0.3234	1.9337
41	0.3285	1.9036
42	0.3274	1.9098
43	0.3279	1.9070
44	0.3279	1.9067
45	0.3276	1.9089
46	0.3270	1.9122
47	0.3275	1.9094
48	0.3282	1.9052
49	0.3285	1.9033

50	0.3277	1.9079
51	0.3287	1.9022
52	0.3275	1.9093
53	0.3278	1.9074
54	0.3281	1.9057
55	0.3272	1.9109
56	0.3278	1.9078
57	0.3275	1.9091
58	0.3280	1.9063
59	0.3279	1.9072
60	0.3273	1.9107
61	0.3278	1.9078
62	0.3279	1.9070
63	0.3270	1.9121
64	0.3273	1.9102
65	0.3272	1.9110
66	0.3263	1.9161
67	0.3271	1.9116
68	0.3266	1.9144
69	0.3271	1.9114
70	0.3269	1.9126
71	0.3271	1.9114
72	0.3266	1.9148
73	0.3261	1.9177
74	0.3251	1.9233
75	0.3255	1.9208
76	0.3269	1.9129
77	0.3261	1.9175
78	0.3252	1.9229
79	0.3244	1.9277
80	0.3255	1.9212
81	0.3245	1.9270
82	0.3243	1.9279
83	0.3244	1.9274

84	0.3242	1.9285
85	0.3232	1.9349
86	0.3234	1.9338
87	0.3245	1.9269
88	0.3233	1.9339
89	0.3237	1.9316
90	0.3247	1.9255
91	0.3242	1.9285
92	0.3259	1.9188
93	0.3252	1.9231
94	0.3241	1.9296
95	0.3241	1.9291
96	0.3242	1.9286
97	0.3240	1.9301
98	0.3240	1.9296
99	0.3250	1.9241

Bisher erschienene Bände der Reihe

Science-Report aus dem Faserinstitut Bremen

ISSN 1611-3861

1	Thomas Körwien	Konfektionstechnisches Verfahren zur Herstellung von endkonturnahen textilen Vorformlingen zur Versteifung von Schalensegmenten ISBN 978-3-8325-0130-3 40.50 EUR
2	Paul Jörn	Entwicklung eines Produktionskonzeptes für rahmenförmige CFK-Strukturen im Flugzeugbau ISBN 978-3-8325-0477-9 40.50 EUR
3	Mircea Calomfirescu	Lamb Waves for Structural Health Monitoring in Viscoelastic Composite Materials ISBN 978-3-8325-1946-9 40.50 EUR
4	Mohammad Mokbul Hossain	Plasma Technology for Deposition and Surface Modification ISBN 978-3-8325-2074-8 37.00 EUR
5	Holger Purol	Entwicklung kontinuierlicher Preformverfahren zur Herstellung gekrümmter CFK-Versteifungsprofile ISBN 978-3-8325-2844-7 45.50 EUR
6	Pierre Zahlen	Beitrag zur kostengünstigen industriellen Fertigung von haupttragenden CFK-Großkomponenten der kommerziellen Luftfahrt mittels Kernverbundbauweise in Harzinfusionstechnologie ISBN 978-3-8325-3329-8 34.50 EUR
7	Konstantin Schubert	Beitrag zur Strukturzustandsüberwachung von faserverstärkten Kunststoffen mit Lamb-Wellen unter veränderlichen Umgebungsbedingungen ISBN 978-3-8325-3529-2 53.00 EUR
8	Christian Brauner	Analysis of process-induced distortions and residual stresses of composite structures ISBN 978-3-8325-3528-5 52.00 EUR
9	Tim Berend Block	Analysis of the mechanical response of impact loaded composite sandwich structures with focus on foam core shear failure ISBN 978-3-8325-3853-8 55.00 EUR

10	Hendrik Mainka	Lignin as an Alternative Precursor for a Sustainable and Cost-Effective Carbon Fiber for the Automotive Industry ISBN 978-3-8325-3972-6 43.00 EUR

Alle erschienenen Bücher können unter der angegebenen ISBN-Nummer direkt online (http://www.logos-verlag.de) oder per Fax (030 - 42 85 10 92) beim Logos Verlag Berlin bestellt werden.